MATHS & STATS

FOR THE LIFE AND MEDICAL SCIENCES

MATHS & STATS

FOR THE LIFE AND MEDICAL SCIENCES

SECOND EDITION

MICHAEL HARRIS

Associate Postgraduate Dean, Severn School of Primary Care, Bristol, UK

GORDON TAYLOR

Reader in Medical Statistics, University of Bath, Bath, UK

JACQUELYN TAYLOR

Private tutor in mathematics and science, Pilton, UK

Scion

Second edition first published 2013

© Scion Publishing Ltd, 2013

First edition published 2005, reprinted 2006, 2007, 2008, 2009, 2010, 2011 and 2012

A CIP catalogue record for this book is available from the British Library.

ISBN 978 1 904842 90 3

Scion Publishing Limited

The Old Hayloft, Vantage Business Park, Bloxham Rd, Banbury, OX16 9UX, UK

www.scionpublishing.com

Important Note from the Publisher

The information contained within this book was obtained by Scion Publishing Limited from sources believed by us to be reliable. However, while every effort has been made to ensure its accuracy, no responsibility for loss or injury whatsoever occasioned to any person acting or refraining from action as a result of information contained herein can be accepted by the authors or publishers.

Typeset by Phoenix Photosetting, Chatham, Kent, UK
Printed in the UK by 4edge Ltd, Hockley, Essex

Contents

Preface

This book is designed for life and medical science students and professionals who need a basic knowledge of mathematics and statistics.

Whether you love or hate maths and stats, you need to have some working knowledge of the subjects if you want to work in the life or medical sciences.

This book assumes that you have nothing more than a very basic maths or stats knowledge. However basic your knowledge, you will find that everything is clearly presented and explained.

A few readers will find some of the sections very simplistic; others will find that some need a lot of concentration. Start with concepts that suit your level of understanding.

All the sections have worked examples, and you can check your understanding of what you have learnt by going through the "test yourself" questions, then comparing your answers with those of the authors.

For this second edition, we have revised and added to the chapters on mathematics. As a result of comments from our readers, we have considerably increased the number of "test yourself" questions and worked answers.

Michael Harris
Gordon Taylor
Jacquelyn Taylor
Somerset, March 2013

About the authors

Dr Michael Harris MB BS FRCGP MMEd is a GP and Associate Postgraduate Dean at the Severn School of Primary Care in Bristol. He has a special interest in the design of educational materials.

Dr Gordon Taylor PhD MSc BSc (Hons) is a Reader in Medical Statistics at the University of Bath. His main role is in the teaching, support and supervision of health-care professionals involved in non-commercial research.

Mrs Jacquelyn Taylor MSc BSc (Hons) PGCE has worked in both secondary and higher education in mathematics and science.

Acknowledgements

We would like to thank all our reviewers, whether expert or enthusiastic amateur.

We are very grateful to Professor Jonathan Sherratt of Heriot-Watt University for his comments and for having given us permission to use some of his material.

Thank you also to our publisher, Dr Jonathan Ray, for his patience and helpful advice.

Finally, special thanks go to Sue Harris for her forbearance and support.

How to use this book

If you want a maths and stats course

- Work through from start to finish for a complete course in the mathematics and statistics relevant to the life and medical sciences.
- The first page starts with the assumption that you want to go right back to basics.
- If you already know some maths or stats, start with concepts that suit your level of understanding.
- Each chapter will build on what you have learnt in previous chapters.
- All the chapters have worked examples that illustrate what you have read. These examples will help you reinforce your learning.
- We have cut down the jargon as much as possible. All new words are put in bold and explained.

If you're in a hurry

- Choose the chapters that are relevant to you. Each chapter is designed so that it can be read in isolation.

If you want a reference book

- You can use this as a reference book. The index is detailed enough for you to find what you want in a hurry.

Test your understanding

- Use the "test yourself" questions at the end of the chapters to check your understanding of what you have just read, then compare your answers with those of the authors.
- You will be able to answer most questions by using mental arithmetic. Some questions will be easier to answer if you use a calculator.

Study advice

- Try not to cover too much at once.
- Go through difficult sections when you are fresh.
- You may need to read some sections a couple of times before the meaning sinks in. You will find that working through the examples helps you to understand the principles.

Handling numbers

This chapter goes right back to basics, with reminders of the principles of handling numbers.

2.1 Factors

The **factors** of a number are all the whole numbers that divide into it without a remalnder.

The numbers 1, 2, 3, 4, 6 and 12 all divide into 12 exactly. These numbers are called the factors of 12.

$$1 \times 12 = 12$$

$$2 \times 6 = 12$$

$$3 \times 4 = 12$$

> **EXAMPLE**
>
> The factors of 15 are 1, 3, 5 and 15.

2.2 Common factors

Common factors are factors that two or more numbers have in common.

> **EXAMPLE**
>
> 1 and 3 are the common factors of 12 and 15.
>
> The **highest common factor** is 3.

2.3 Use of brackets

We use **brackets** to change the order of mathematical calculation and ensure correct interpretation of mathematical **expressions**.

> **EXAMPLE**
>
> The expression
>
> $$3 \times 8 - 5$$

is calculated as follows:

$3 \times 8 = 24$; subtract 5; answer: 19.

Putting brackets round the $8 - 5$ expression changes the order of calculation: the contents of brackets need to be worked out *before* doing the rest of a calculation.

$3(8 - 5) = 3 \times 3 = 9$

Note that when we write "3 times $(8 - 5)$" we don't need to write the "$\times$" sign.

2.4 BIDMAS

Calculations in a mathematical expression should be in **BIDMAS** order:

- Brackets
- Indices: this relates to powers of numbers
- Division
- Multiplication
- Addition
- Subtraction

EXAMPLE

To calculate

$9 + 6 - 8(7 + 5) \div 4$

First we work out the sum in brackets, giving:

$9 + 6 - 8(12) \div 4$

After the division we get:

$9 + 6 - 8 \times 3$

After multiplying:

$9 + 6 - 24$

After the addition:

$15 - 24$

Finally, the subtraction gives:

-9

2.5 The multiplication sign

The only sign that can be omitted in an equation is $\times$, the multiplication sign. Sometimes we use a dot $\cdot$ instead of the $\times$ sign. This is usually to prevent any confusion between multiplication and an algebraic x.

> **EXAMPLE**
>
> $3 \times x$ can be written as $3x$ and also as $3 \cdot x$

2.6 Absolute values

Absolute value bars around a number turn it into a positive value. Absolute value does nothing to a positive number or zero.

> **EXAMPLE**
>
> $|-3| = 3$
>
> $|3| = 3$

2.7 Prime numbers

A whole number with only two factors, 1 and itself, is called a **prime number**.

> **EXAMPLE**
>
> 7 has only two factors, 1 and 7, so it is a prime number.

2.8 Prime factors

Prime factors are factors that are also prime numbers. Any number can be written in terms of its prime factors.

A **factor tree** is a diagram that breaks down a number by dividing it by its factors until all the numbers left are prime.

> **EXAMPLE**
>
> ```
>     42
>     / \
>    7   6
>    /   / \
>   7   3   2
> ```
>
> The prime factors of 42 are 7, 3 and 2.

2.9 Square numbers

Square numbers are formed by multiplying a whole number by itself, for example:

9 is 3×3, 25 is 5×5

3×3 can be written as 3^2, and can be spoken as "3 squared" or "3 to the power of 2".

When two negative numbers are multiplied the answer is positive, so the square of a negative number is positive.

EXAMPLES

$(-3)^2 = -3 \times -3 = 9$

$5 \times 5 = 5^2 = 25 = $ "5 squared", or "5 to the power of 2".

A **quadrat** is an area used as a sample unit, typically with a size of 1 by 1 m.

So, a 5 by 5 metre square area of a field could contain 25 separate quadrats.

2.10 Square roots

Because $9 = 3 \times 3$, 3 is called the **square root** of 9. The square root of a number is written using the $\sqrt{}$ symbol, so $\sqrt{9} = 3$.

However, we know that $(-3)^2$ also equals 9, so $\sqrt{9}$ could also be −3.

The square root of any number can be positive or negative, so $\sqrt{16} = \pm 4$.

The $\pm$ symbol means "plus or minus".

2.11 Cube numbers

Cube numbers are formed by multiplying a number by itself and then by itself again.

EXAMPLE

$5 \times 5 \times 5 = 5^3 = 125 = $ "5 cubed", or "5 to the power of 3".

Thus a block of plant tissue 5 by 5 by 5 millimetres is $125 \, \text{mm}^3$.

2.12 Cube roots

Because $27 = 3 \times 3 \times 3$, 3 is the **cube root** of 27.

This is written as $\sqrt[3]{27} = 3$.

Test yourself

The answers are given on page 207.

Question 2.1
What are the factors of 18, 21 and 24?
What is their highest common factor?

Question 2.2
Calculate
$7(4 + 3)(5 - 2)$

Question 2.3
Work out
$16(9 \div 3 + 1) - 10 \div 5$

Question 2.4
Which of these are prime numbers: 21, 22, 23?

Question 2.5
What is the surface area of a 7 by 7 m square sample of a field?

Question 2.6
What is the length of a side of a square skin biopsy sample that is $64\,mm^2$?

Question 2.7
A square nicotine patch releases 0.6 mg of nicotine per cm^2 in a 24 hour period. A patient needs 15 mg in 24 hours. What area of patch does the patient need? What dimensions are needed?

Question 2.8
What is the volume of a 40 by 40 by 40 mm cube of soil sample?

Question 2.9
Gold-coated metal perforated cubes are being trialled as a delivery system for medications. The boxes have a side length of 3.4 nm. What is the volume of these cubes?

Question 2.10
A cube of tissue from a biopsy is $64\,mm^3$. What are its dimensions?

Working with fractions

As soon as we work with anything other than whole numbers, we need to use fractions.

3.1 Fractions

The **fraction** $\frac{3}{5}$ (often presented as 3/5) means 3 parts out of 5. The top number in a fraction is known as the **numerator**, the bottom number is called the **denominator**.

To calculate a fraction of an amount, multiply by the numerator and divide by the denominator.

> **EXAMPLE**
>
> $\frac{3}{5}$ of $20 = (3 \times 20) \div 5 = 60 \div 5 = 12$

3.2 Simplifying fractions

Fractions can be **simplified** if the numerator and denominator have a common factor.

With fractions, whatever we do to the numerator, we also have to do to the denominator.

> **EXAMPLE**
>
> To simplify 12/15, the 12 and the 15 have a common factor of 3, so we can divide the numerator and denominator by 3.
>
> $$\frac{12}{15} = \frac{12 \div 3}{15 \div 3} = \frac{4}{5}$$
>
> Where there is no common factor, the fraction is already in its simplest form.

3.3 Reciprocals

The **reciprocal** of a number or a mathematical expression is 1 divided by that number or expression.

To get the reciprocal of a fraction, flip it upside down.

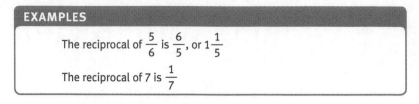

EXAMPLES

The reciprocal of $\dfrac{5}{6}$ is $\dfrac{6}{5}$, or $1\dfrac{1}{5}$

The reciprocal of 7 is $\dfrac{1}{7}$

3.4 Multiplying fractions

To **multiply fractions**, multiply straight across the top, and straight across the bottom.

EXAMPLE

$$\frac{3}{4} \times \frac{5}{6} = \frac{3 \times 5}{4 \times 6} = \frac{15}{24} = \frac{5}{8}$$

3.5 Dividing fractions

To **divide a fraction** by another, flip the second fraction (i.e. take its reciprocal) and multiply the two fractions.

EXAMPLE

$$\frac{3}{4} \div \frac{5}{6} = \frac{3}{4} \times \frac{6}{5} = \frac{3 \times 6}{4 \times 5} = \frac{18}{20} = \frac{9}{10}$$

3.6 Adding fractions

Where the denominator is the same in both fractions, i.e. where there is a **common denominator**, add across the top.

EXAMPLE

$$\frac{3}{7} + \frac{2}{7} = \frac{3+2}{7} = \frac{5}{7}$$

Where there is *no* common denominator, the simplest way is to convert the fractions so that they have the **lowest common denominator,** and then add across the top.

The lowest common denominator is the smallest number that has all the denominators as factors.

> **EXAMPLE**
>
> $$\frac{2}{3} + \frac{4}{5}$$
>
> The lowest common denominator is 15, as it is the smallest number that has 3 and 5 as factors.
>
> To convert 2/3 to have 15 as a denominator, we need to multiply the numerator and denominator by 5.
>
> To convert 4/5 to have 15 as a denominator, we need to multiply the numerator and denominator by 3.
>
> $$\frac{2 \times 5}{3 \times 5} + \frac{4 \times 3}{5 \times 3} = \frac{10}{15} + \frac{12}{15} = \frac{10 + 12}{15} = \frac{22}{15} = 1\frac{7}{15}$$

3.7 Subtracting fractions

The process is analogous to that for adding fractions.

> **EXAMPLE**
>
> $$\frac{3}{7} - \frac{2}{7} = \frac{3 - 2}{7} = \frac{1}{7}$$

Again, where there is no common denominator, convert each fraction to the lowest common denominator and subtract across the top.

3.8 Changing fractions to decimals

To change a fraction into its **decimal** equivalent, divide the numerator by the denominator.

> **EXAMPLE**
>
> $$\frac{1}{2} = 1 \div 2 = 0.5$$

3.9 Recurring decimals

A fraction like 1/3, when written as a decimal, becomes 0.333333... where the threes continue forever. We call these **recurring decimals**. A dot is used above the recurring digit as shorthand.

> **EXAMPLE**
>
> $$\frac{2}{3} = 0.666666... = 0.\dot{6}$$

Where a series of digits is repeated again and again, dots are placed over the first and last digits of the series.

EXAMPLE

$$\frac{1}{7} = 0.142857142857142857... = 0.\dot{1}4285\dot{7}$$

Test yourself

The answers are given on page 207.

Question 3.1
Calculate $\frac{5}{6}$ of 72.

Question 3.2
Simplify $\frac{20}{24}$

Question 3.3
Give the reciprocal of $\frac{24}{28}$

Question 3.4
Multiply $\frac{2}{5}$ by $\frac{9}{10}$

Question 3.5
Divide $\frac{2}{5}$ by $\frac{9}{10}$

Question 3.6
Add $\frac{6}{7}$ to $\frac{9}{14}$

Question 3.7
Subtract $\frac{7}{12}$ from $1\frac{3}{8}$

Question 3.8
What is the decimal equivalent of $1\frac{5}{8}$?

Question 3.9
A child needs to be given one-fifth of the daily adult dose of a medicine in a 24 hour period. The medication needs to be administered three times a day. What fraction of the adult dose needs to be administered each time?

Question 3.10
An adult human male's blood count results are given as:

Red blood cells	3 960 000 per mm³ of blood
White blood cells	10 000 per mm³ of blood
Platelets	430 000 per mm³ of blood

a) What fraction of the total count is made up of platelets?
b) As a decimal, what proportion of the total count is made up of red blood cells?

Question 3.11
An 80 year old patient sleeps for 7 hours each night, of which 2 hours are REM (rapid eye movement) sleep.
a) What fraction of the whole day does the patient spend asleep?
b) What fraction of her sleep is spent in REM sleep?
c) If this were typical of her life-long sleep pattern, how much of her life has she spent asleep? Give your answer to the nearest month.

Percentages

A percentage is another way of describing a fraction and it can be easier to visualise.

4.1 Percentages

A **percentage** is a fraction out of 100.

$$15\% \text{ is the same as } \frac{15}{100}$$

So, calculating 15% of a number is the same as calculating 15/100 of the number.

It is also the same as multiplying the number by 0.15, as $0.15 = 15 \div 100$.

> **EXAMPLE**
>
> $$15\% \text{ of } 480 = \left(\frac{15}{100}\right) \times 480 = 72$$
>
> Also, 15% of $480 = 0.15 \times 480 = 72$

4.2 Converting decimals into percentages

To change a decimal into a percentage, multiply it by 100.

> **EXAMPLE**
>
> $$0.05 = (100 \times 0.05)\% = 5\%$$

4.3 Calculating percentages using decimals

To calculate a percentage increase or decrease, convert the percentage to decimals.

> **EXAMPLE**
>
> A stalk of common wheat, *Triticum aestivum*, measures 625 mm in height. In 1 week it grows by 12%.
>
> An increase *of* 12% is the same as an increase *to* 112%; multiplying a number by 112% is the same as multiplying it by 1.12.
>
> 112% of $625 = 1.12 \times 625 = 700$
>
> So, after 1 week the stalk has grown to 700 mm.

When a number increases (or decreases), we can calculate the increase (or decrease) as a percentage of the original number.

Work out the change as a fraction of the original number, then convert it to decimals. Multiply this by 100 to get the percentage.

EXAMPLE

The weight of a baby has increased from 1.3 kg to 1.56 kg.

It has therefore increased by 0.26 kg, or $\dfrac{0.26}{1.30}$ of the original weight.

$$\dfrac{0.26}{1.30} \times 100 = 20\%$$

So, the baby has gained 20% in weight.

4.4 Tabulating data using percentages

We use percentages when **tabulating data** to give a scale on which to assess or compare the data.

EXAMPLE

We can use a table to compare data for body and tail length in 10 mice.

Table comparing body and tail length in 10 mice		
Body length (mm)	Tail length (mm)	$\dfrac{\text{Tail length}}{\text{Body length}}$ (to nearest %)
92	31	34
97	32	33
96	35	36
99	36	36
100	40	40
111	43	39
109	44	40
115	49	43
120	49	41
122	52	43

Note how tabulating the percentages makes the trend clear: in this group of mice, longer mice have proportionately longer tails.

We can also tabulate **frequencies** (the number of times events occur) and compare them with percentages.

EXAMPLE

We wish to compare the ages of 80 patients referred for heart transplantation.

Table comparing ages of 80 patients referred for heart transplantation		
Years	Frequency	Percentage
0–9	2	2.5
10–19	5	6.25
20–29	6	7.5
30–39	14	17.5
40–49	21	26.25
50–59	20	25
≥60	12	15
Totals	80	100

The first column gives age in 10-year ranges.

The ≥ symbol means "more than or equal to", in this case "more than or equal to 60 years old".

The second column gives the frequency, i.e. the number of patients in each 10-year range.

The last column gives the percentage of patients in each age range. For example, in the 30–39 year range there were 14 patients and we know the ages of 80 patients, so:

$$\frac{14}{80} \times 100 = 17.5\%$$

Take care when interpreting percentages, though.

To say that 50% of a sample meets certain criteria when there are only four subjects in the sample is clearly not providing the same level of information as 50% of a sample based on 400 subjects.

So, percentages should be used as an additional help when interpreting data, rather than replacing the actual data.

Test yourself

The answers are given on pages 207–8.

Question 4.1
A research subject has a mass of 72 kg. A test shows that 43.2 kg of this is water. What percentage of the subject's total mass is water?

Question 4.2
Drying a 375 g sample of soil has reduced the mass by 40%. What mass of water was there in the sample?

Question 4.3
A patient's peak flow rate (a measure of respiratory airflow) is 400 litres per minute during an attack of asthma.
Twenty minutes after treatment, his peak flow has increased to 560 litres per minute. What percentage increase is this?

Question 4.4
The mean waist size of a sample of patients newly diagnosed with diabetes was 120 cm. After they have been on insulin for a year, their mean waist size has increased to 129 cm. By what percentage has their waist size increased?

Question 4.5
The mean mass of leaf litter per square metre in an area of woodland is 900 g. The mass reduces by 18% in a month. What is the resulting mean mass?

Question 4.6
In 100ml of cell culture at 37°C, the concentration of *E. coli* is 24 million cells per ml immediately after inoculation.

Three hours later the cell concentration has increased to 912 million cells per ml.
What is the percentage increase?

Question 4.7
In 1 year, 1.6 million (1 600 000) women were invited to have breast cancer screening and 75% of the women invited were actually screened. Of those screened, 1% were found to have breast cancer. However, screening only finds 80% of the breast cancers that are present at the time of screening.
a) Of those invited, how many women were screened?
b) How many breast cancers were identified by screening?
c) Of those screened, how many women had breast cancer that was not identified?

Question 4.8
A doctor asks a pharmacist to prepare 0.25 g of a drug. This drug is available in a 0.5% w/v solution (0.5 g per 100 ml). How many ml does the patient need?

Question 4.9
A man in a weight loss programme reduced his body mass by 20% in a year. However, in the subsequent year he gained 20%. He weighed 150 kg when he first entered the programme. How much did he weigh at the end of the second year?

05 Powers

Powers are used throughout the life and medical sciences, whether for describing very large and very small numbers, to describe exponential and other relationships, or for their use in calculus and statistical analysis.

5.1 Indices

The **power** of a number is the same as the **index** of a number (plural: **indices**).

3^2 is known as "3 to the power of 2", or "3 squared",

3^3 is "3 to the power of 3", or "3 cubed",

$3 \times 3 \times 3 \times 3$ is 3^4, or "3 to the power 4",

and so on.

The number being multiplied is known as the **base**. In the examples above, the base is 3.

5.2 Powers of 10

We use **powers of 10** to describe very large and very small numbers.

Table of powers of 10			
Power of 10	Spoken as	Calculated by	Ordinary form
10^4	"10 to the power of 4"	$10 \times 10 \times 10 \times 10$	10 000
10^3	"10 to the power of 3" or "10 cubed"	$10 \times 10 \times 10$	1000
10^2	"10 to the power of 2" or "10 squared"	10×10	100
10^1	"10 to the power of 1"	10	10
10^0	"10 to the power of zero"	$\dfrac{10}{10}$	1
10^{-1}	"10 to the power of minus 1"	$\dfrac{1}{10}$	$\dfrac{1}{10}$ (or 0.1)
10^{-2}	"10 to the power of minus 2"	$\dfrac{1}{10 \times 10}$	$\dfrac{1}{100}$ (or 0.01)

We also use powers of 10 to describe other large numbers:

$$2\,380\,000 = 2.38 \times 1\,000\,000 = 2.38 \times 10^6$$

The $2\,380\,000$ format is known as **ordinary form**.

The 2.38×10^6 format, where there is only one digit before the decimal point, is known as **standard form**.

Note that, for each move of the decimal point to the left, the power increases by 1.

In this example, to get standard form the decimal point has been moved six places to the left, so it results in 10 to the power of 6.

Small numbers can be written in standard form as well:

$$0.0056 = 5.6 \times 0.001 = 5.6 \times 10^{-3}$$

Here, for each move of the decimal point to the right the power reduces by 1.

So, to get standard form the decimal point has been moved three places to the right, resulting in 10 to the power of −3.

5.3 Multiplying and dividing powers with the same base

When we multiply numbers with powers, we have to *add* the powers.

EXAMPLE

$$3^3 \times 3^2 = 3^{3+2} = 3^5 = 3 \times 3 \times 3 \times 3 \times 3 = 243$$

$$10^4 \times 10^2 = 10^{4+2} = 10^6 = 10 \times 10 \times 10 \times 10 \times 10 \times 10 = 1\,000\,000$$

Similarly, to divide a number with a power by another, we *subtract* the powers.

EXAMPLE

$$3^6 \div 3^2 = 3^{6-2} = 3^4 = 3 \times 3 \times 3 \times 3 = 81$$

$$10^7 \div 10^3 = 10^{7-3} = 10^4 = 10 \times 10 \times 10 \times 10 = 10\,000$$

5.4 Multiplying or dividing numbers in standard form

When multiplying or dividing numbers in standard form, group the numbers together, and group the powers of 10 together.

EXAMPLE

A sample of water has 2200 (2.2×10^3) bacteria per litre. To calculate how many bacteria there are in 36 000 litres (3.6×10^4 l) of water, we need to multiply 3.6×10^4 by 2.2×10^3.

Grouping the numbers (3.6×2.2) and also the powers ($10^4 \times 10^3$) together, we get:

$$(3.6 \times 2.2)(10^4 \times 10^3) = 7.92 \times 10^{4+3} = 7.92 \times 10^7$$

Note that (3.6×2.2) ($10^4 \times 10^3$) means (3.6×2.2) $\times$ ($10^4 \times 10^3$).

So there are 7.92×10^7 bacteria in the 36 000 litres of water.

5.5 Addition and subtraction of numbers in standard form

To add or subtract numbers In standard form, If the powers are the same:

- the two numbers can be added or subtracted;
- the power remains the same.

EXAMPLE

Two samples of tissue weigh 2.3×10^{-3} and 5.6×10^{-3} kg.

Their total mass is:

$$(2.3 + 5.6) \times 10^{-3} = 7.9 \times 10^{-3} \text{ kg}$$

If the powers are different, convert one away from standard form so that the powers are the same and then add or subtract the numbers, leaving the power the same. This can then be converted back to standard form.

EXAMPLE

Two samples of seawater are 4.41×10^7 and 7.9×10^5 mm^3.

It doesn't matter which sample is converted away from standard form, the end result will be the same.

If we convert the first sample,

$$4.41 \times 10^7 = 441 \times 10^5$$

Their total volume is therefore:

$$(441 + 7.9) \times 10^5 = 448.9 \times 10^5 \text{ mm}^3$$

Converting back to standard form gives:

$$4.489 \times 10^7 \text{ mm}^3$$

Test yourself

The answers are given on page 208.

Question 5.1
An onion leaf epidermal cell, *Allium cepa*, is 0.00045 m long. Give this in standard form.

Question 5.2
Human genomic DNA is made up of approximately 3×10^9 base pairs. Give this in ordinary form.

Question 5.3
A vitamin tablet contains 800 µg of vitamin A, 5 µg of vitamin D and 12 mg of vitamin E. Compare the relative amounts by writing the weights in grams in standard form, and also in ordinary form. Note that 1 mg = 1×10^{-3} g and 1 µg = 1×10^{-6} g.
Which vitamin is present in the highest dose? Which is present in the lowest dose?

Question 5.4
It is estimated that in a rural area there is a mean of 150 people per square kilometre. Using standard form, calculate how many people will there be in a square plot of 40 by 40 km.

Question 5.5
Some ocean phytoplankton are found to be 0.015 mm in length, while a species of whale has a mean length of 30 m.
a) Rewrite these sizes in metres in standard form.
b) How many times longer are the whales than the phytoplankton?

Circles and spheres

The mathematical formulae relating to circles and spheres can be helpful in many scientific calculations.

This diagram shows how the words "radius", "diameter" and "circumference" relate to a circle.

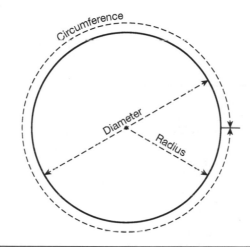

Elements of a circle

6.1 Pi

All the formulae for circles and spheres relate to the mathematical constant π. You may see π written as **pi,** and it is pronounced "pie".

π is an infinitely long number: π is approximately equal to 3.1416.

6.2 Formulae for circles and spheres

For circles and spheres of radius r:

Circumference of a circle	$C = 2\pi r$
Area of a circle	$A = \pi r^2$
Volume of a sphere	$V = \dfrac{4}{3}\pi r^3$
Surface area of a sphere	$A = 4\pi r^2$

21

6.3 Formulae for cylinders and cones

For cylinders and cones of radius r and height h:

Volume of a cylinder $\qquad V = \pi r^2 h$

Volume of a cone $\qquad V = \dfrac{1}{3} \pi r^2 h$

> **EXAMPLE**
>
> A glass cylinder of radius 120 mm contains solution up to a height of 85 mm. We wish to know the volume of solution.
>
> Substituting the values into the formula for the volume of a cylinder:
>
> $V = \pi r^2 h = 3.1416 \times 120^2 \times 85$ which is approximately equal to 3.8×10^6 mm^3.

Test yourself

The answers are given on page 208.

Question 6.1
One drop of the antiseptic chlorhexidine gluconate 2% in 70% isopropyl alcohol on the skin spreads to form a circle of diameter 2 mm.
a) What area would be covered if three drops were applied? Give your answer to three significant figures.
b) The antiseptic penetrates 0.5 mm into the skin. What volume of skin is treated by three drops?

Question 6.2
An egg yolk to be used for cell culture is 24 mm diameter. Assuming it is spherical, calculate its volume.

07 Approximation and errors

We sometimes need to **approximate** numbers and give the **degree of accuracy**.

7.1 Approximation

One way of approximating is to give the **nearest whole number**.

32.543716 is nearer 33 than 32, so it is "33 to the nearest whole number".

Another way of giving the degree of accuracy when approximating is to state the number of **decimal places**.

32.543716 is 32.54 to two decimal places, and 32.5437 to four decimal places.

A third way of giving the degree of accuracy when approximating is to give the number of **significant figures**, i.e. the number of digits quoted.

32.543716 is 32.54 correct to four significant figures.

> **EXAMPLE**
>
> 28365 is 28000 correct to two significant figures.

7.2 Significant figures and handling zeros

When there are zeros *within* a number, the zeros are considered as significant figures.

> **EXAMPLE**
>
> 10.54 counts as four significant figures.

If the *last* figures in a *whole* number are zeros, the zeros are not counted as significant figures.

> **EXAMPLE**
>
> 6754000 counts as four significant figures.

23

If the zeros are *before* a decimal number, the zeros are not counted as significant figures.

> **EXAMPLE**
>
> 0.0004832 counts as four significant figures.

However, zeros *after* a decimal number are counted as significant figures.

> **EXAMPLE**
>
> 0.8760 counts as four significant figures.

7.3 Rounding numbers

When approximating numbers, the last significant digit stays as it is, i.e. the number is "rounded down", if the next digit is below 5.

> **EXAMPLE**
>
> 6340 is 6300 to two significant figures.

The last significant digit is "rounded up" if the next digit is 5 or above.

> **EXAMPLE**
>
> 6360 is 6400 to two significant figures.

Note that when the next digit is exactly 5, the common convention is to round the last significant digit up.

> **EXAMPLE**
>
> 6350 is 6400 to two significant figures.

7.4 Choosing the number of significant figures

When two or more measurements are taken, the result is only as reliable as the least reliable value.

So, to work out how many significant figures to use, use the same number of significant figures as the least precise value that the result was derived from.

If you need to adjust the number of significant figures, always do so at the end of your calculations.

> **EXAMPLE**
>
> An animal is noted to have run 47.81 metres in 8.5 seconds. We want to know its velocity.
>
> The least precise value was the time, which was measured to two significant figures. So, the velocity can also only be given to two significant figures.
>
> $$\frac{47.81}{8.5} = 5.6247 = 5.6\,\mathrm{m\,s^{-1}} \text{ to two significant figures.}$$
>
> Note that "metres per second" can be written as $\mathrm{m\,s^{-1}}$ or as m/s.

7.5 Errors

Where we use approximations, there will be some **errors** in the calculations.

If the height of a plant is given as 4 m, the lack of a decimal place implies that the measurement is correct to the nearest metre. There is an error of ± 0.5 m, so the actual height could be anywhere between 3.5 and just below 4.5 m (remember that a height of exactly 4.5 m would have been rounded up to 5 m).

If the plant's height is given as 4.29 m, the error is ± 0.005 m, and the actual height could be anywhere between 4.285 and just below 4.295 m.

7.6 Precision and accuracy

Measuring instruments may be precise, in that they can give results to many significant figures, but inaccurate, in that they may be incorrectly calibrated.

> **EXAMPLE**
>
> A pH meter may measure pH to three decimal places of precision. However, if it has been set up incorrectly it will always give an inaccurate value.

Test yourself

The answers are given on page 208.

Question 7.1
The height of a child is 1.050 m. How many significant figures is this?

Question 7.2
58.44 g of NaCl is dissolved in 0.137 m^3 of water. Use a calculator to work out the concentration, and state your answer to the appropriate number of significant figures.

Question 7.3
The mass of a hen's egg is given as 56 g to the nearest gram. Within what range could the actual mass be?

Question 7.4
Calibration of equipment used to measure haemoglobin levels in blood shows that its readings can vary up to $\pm$ 1.3 g dl^{-1}. The equipment gives an adult male patient's haemoglobin as 14.9 g dl^{-1}. The normal range for haemoglobin level in adult men is 13.5–18.0 g dl^{-1}. Can you be sure that this patient's haemoglobin levels are in the normal range?

Question 7.5
An *E. coli* bacterium is studied under an electron microscope and found to measure 295.638×10^{-9} m. Write this to 4 significant figures:
a) in standard form.
b) in ordinary form.

Introduction to graphs

One way to present data is in the form of a table. However, often we can understand and interpret data more easily by plotting them on a graph.

Graphs also help to define the relationship between two variables.

8.1 The *x*- and *y*-axes

To compare two variables we use a two-dimensional graph. It uses an **x-axis** and a **y-axis**.

The horizontal axis is known as the *x*-axis, and the vertical axis is called the *y*-axis.

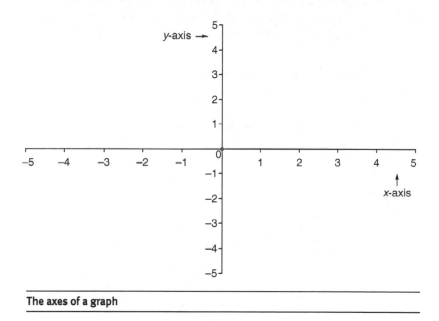

The axes of a graph

The variable that is "controlled" by the researcher is usually plotted on the *x*-axis. We call this the **independent variable,** as it is not dependent on the other variable, and can therefore be determined by the researcher.

We plot the **dependent variable** on the *y*-axis. This is the variable which can be determined or predicted if the *x* value is known.

When measuring a reaction over time, time is plotted on the *x*-axis as the researcher will decide at what times to measure the reaction.

8.2 Plotting values on a graph

Say that we want to **plot** the values in the table below on a graph.

Table of *x* and *y* values				
Value for *x*	0	2	4	6
Value for *y*	1	3	5	7

First we need to draw the *x*- and *y*-axes to a suitable scale. In this example *x*- and *y*-axes can be drawn from 0 to 8.

We then plot the values.

For example, with the first pair of values, where $x = 0$ and $y = 1$, we find 0 on the *x*-axis and then move vertically up until we reach the value of 1 on the *y*-axis, and mark that point.

While computer software, books and journals will use dots, when drawing a graph by hand it is best to use a cross, as this pinpoints the exact place with more accuracy than a dot.

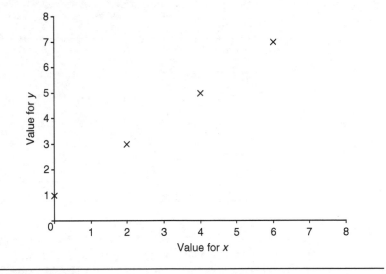

Plot of *x* and *y* values

8.3 Co-ordinates

We can define a point on a graph by its **co-ordinates**.

A co-ordinate is written like this:

(*x*-value, *y*-value)

Therefore the co-ordinates of the points in the plot above are:

(0,1) (2,3) (4,5) and (6,7)

The co-ordinate (0,0) is known as the **origin** of a graph.

8.4 Direct proportion

Variables are said to be in **direct proportion** if:

* when one variable is zero, the other is also zero;
* when one variable changes, the other changes in the same ratio.

> **EXAMPLE**
>
> The number of plankton in a water sample is in direct proportion to the size of the water sample. Doubling the volume of water doubles the number of plankton. If there is no water, there are no plankton.

We use the symbol $\propto$ to indicate direct proportion. So, $x \propto y$ indicates that the value of variable x is directly proportional to the value of variable y.

This can be shown in the form of a graph.

> **EXAMPLE**
>
> This linear graph (straight line graph) shows how the number of plankton in a sample of seawater relates to the size of the sample.

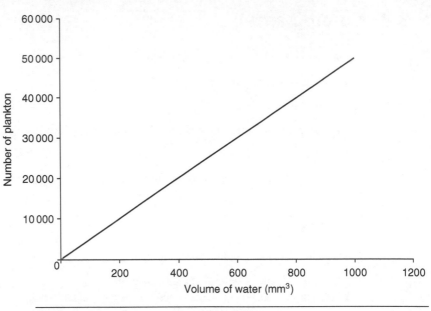

Graph of number of plankton per volume of water

A **linear graph** (straight line graph) that goes through the origin can be written as an equation:

$$y = mx$$

where m is the gradient (or slope) of the graph.

When two variables are in direct proportion, if we know both variables we can calculate the gradient of the graph.

If we know one variable and the gradient, we can calculate the value of the other variable.

Test yourself

The answers are given on page 209.

Question 8.1
In a maze-learning experiment, the number of errors made by a rat in a maze is tabulated against the number of trips.
Plot these values on a graph.

Table of number of errors made by a rat in a maze						
Trip number	1	2	3	4	5	6
Error score	31	18	15	6	7	3

Question 8.2
The energy in a particle is in direct proportion to the temperature of the particle.
The gas constant (R) is approximately 8 units of energy per degree of temperature for a unit of gas. Avogadro calculated that there are approximately 6 particles per unit of gas. Scientists use a constant of proportionality (K) for this relationship, whose units are energy per degree of temperature per particle. This is known as the Stefan–Boltzmann constant. Given that $R = K(N_A)$, calculate K.

Question 8.3
The "linear no-threshold" model is used in radiation protection. It states that the long-term biological damage (number of additional cancer cases) caused by ionising radiation is directly proportional to the radiation dose.
A particular dose of radiation is found to produce one additional case of thyroid cancer in every thousand people exposed. How does this change if a population is exposed to only one thousandth of this dose?

The gradient of a graph

The gradient of a graph describes the steepness of the graph.

9.1 The gradient of a straight line

To calculate the gradient of a graph, we take a section of the graph that we have drawn, then divide the number of units that the graph has moved up the y-axis by the number of units it has moved along the x-axis.

The **gradient** (sometimes known as the **slope**) is therefore $\dfrac{\text{Change in } y}{\text{Change in } x}$.

> **EXAMPLE**
>
> In this graph, for every two units up the y-axis, the line goes one unit along the x-axis.

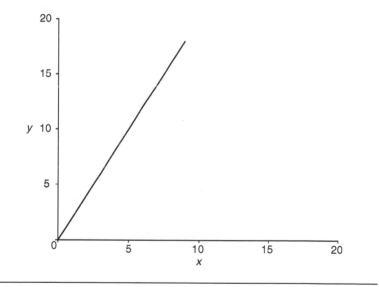

Graph of $y = 2x$

The gradient $\dfrac{\text{Change in } y}{\text{Change in } x}$ is therefore $\dfrac{2}{1}$, or 2, so the equation for the graph is $y = 2x$.

The steeper the graph, the larger the value for the gradient.

This is the graph for $y = \dfrac{x}{3}$ plotted using the same scale axes as the previous graph.

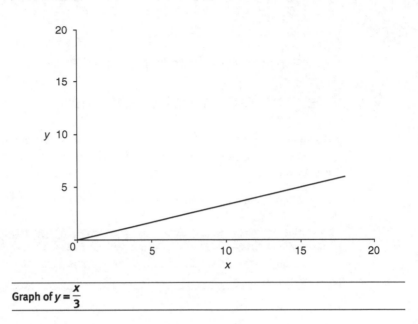

Graph of $y = \dfrac{x}{3}$

The gradient can be given as a decimal. The gradient of this graph is $0.\dot{3}\dot{3}$.

9.2 The formula for the gradient

The formula that describes the gradient

$$\frac{\text{Change in } y}{\text{Change in } x}$$

is

$$\frac{y_2 - y_1}{x_2 - x_1}$$

EXAMPLE

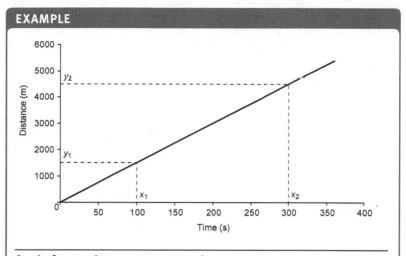

Graph of a car going at a constant speed

This graph shows the distance that a car travels in any given time.

Between 100 and 300 seconds, it has travelled from 1500 to 4500 m.

$$\text{Gradient} = \frac{y_2 - y_1}{x_2 - x_1} = \frac{4500 - 1500}{300 - 100} = \frac{3000}{200}\,\text{m s}^{-1}$$

So, the gradient of the line is 15, making the speed of the car $15\,\text{m s}^{-1}$.

The equation for the graph is therefore $y = 15x$

9.3 The symbol for the gradient

For a straight-line graph, the gradient is symbolised by the constant m.

$$\text{Gradient} = \frac{\text{Change in } y}{\text{Change in } x} = m$$

m is the **rate of change** of y with respect to x.

9.4 Negative gradients

In this graph, for each unit along the *x*-axis, the graph goes two units *down* the *y*-axis:

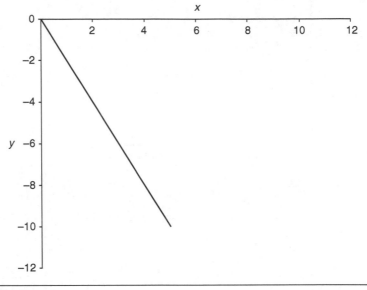

Graph of y = −2x

The gradient is therefore −2/1, or −2, so the equation for the graph is *y* = −2*x*.

So, a line that runs up to the right has a *positive* gradient ...

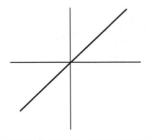

Graph with positive gradient

... while one that runs down to the right has a *negative* gradient:

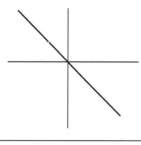

Graph with negative gradient

9.5 Graphs that don't go through the origin

The next graph has a gradient of 4 but the line doesn't go through the origin, i.e. it does not go through (0,0).

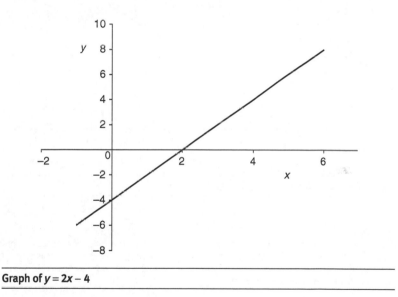

Graph of $y = 2x - 4$

The equation for a graph that does not go through the origin is $y = mx + c$, where m is the gradient (as before), and c is the point at which the line crosses the y-axis (i.e. when $x = 0$ then $y = c$). This point is called the "y-intercept". Note that the x and y axes have different scales.

In this case, the gradient is 2, and the line crosses the y-axis at −4, so the equation for this line is $y = 2x - 4$.

9.6   Linear equations

The equation $y = mx + c$ is called **linear** as it describes a straight-line graph: there is a straight-line relationship between the two variables x and y.

Test yourself

The answers are given on pages 209–10.

Question 9.1
The following graph shows the distance that a cyclist covers over 50 minutes. Calculate the speed that the cyclist is going in kilometres per hour.

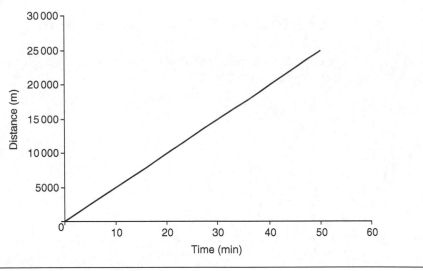

Graph of distance covered by cyclist against time

Question 9.2
A baby girl is 500 mm long when born. She grows 10 mm per week. Give the equation for this relationship and draw a graph that shows how she grows over the first 6 weeks.

Question 9.3
A yew tree, *Taxus baccata*, was planted as a sapling. One year later it is found to be 200 mm high. Two years after that it is found to be 400 mm high. State this growth as an equation and, assuming that the rate of growth is constant, calculate how high it was when planted.

Question 9.4

The relationship between algal uptake of nitrogen (measured as the percentage increase of tissue nitrogen) and the time exposed to nutrient-rich water is shown in the graph below.

What is the mathematical relationship between the percentage increase of nitrogen in dry tissue and the exposure time for the algae?

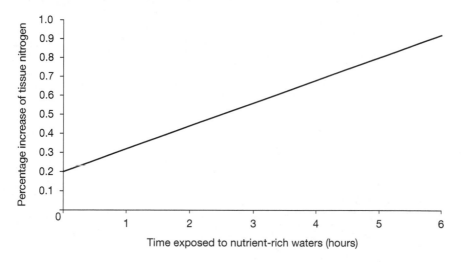

Question 9.5

The relationship between the heights of cotton plants and soil nitrogen rates is shown in the graph below.

a) State the relationship mathematically, i.e. write the equation of the graph.
b) If there are 100 kg of nitrogen per hectare, how tall will the cotton plants be?

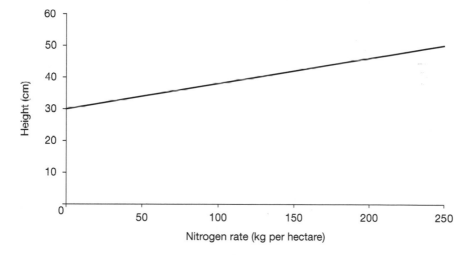

Algebra

Algebra is the area of mathematics that uses symbols to represent numbers. It lets us investigate relationships between quantities.

We need to be competent in algebra to be able to handle ("manipulate") the many equations and formulae used in the life and medical sciences.

10.1 Using symbols

In arithmetic, 2, 5, 7, 9, 10, etc. each have a fixed value.

In algebra, a, b, c, x, y, z, etc. stand for *any* values.

For instance, we may use the letter "x" as a symbol to represent an unknown, variable number. We could equally well use a, b, c or any other letter as a symbol.

10.2 Simplifying expressions

Collecting algebraic terms together helps to **simplify** expressions.

> **EXAMPLES**
>
> $4a + 3a$ can be simplified to $7a$.
>
> $4a + 6b + 3a + b$ can be simplified to $7a + 7b$. This can be further simplified to $7(a + b)$.

Identical powers can be collected together.

> **EXAMPLE**
>
> $2a^2$ is the same as $a^2 + a^2$.
>
> $3a^2$ is the same as $a^2 + a^2 + a^2$.
>
> So, $3a^2$ plus $2a^2$ is the same as $a^2 + a^2 + a^2 + a^2 + a^2$, which can be simplified to $5a^2$.
>
> $4a^4$ plus $5a^4$ can be simplified to $9a^4$.

Different powers cannot be collected together.

> **EXAMPLE**
>
> $2a^2$ is the same as $a^2 + a^2$.
>
> $3a^4$ is the same as $a^4 + a^4 + a^4$.
>
> So, $3a^4$ plus $2a^2$ becomes $a^4 + a^4 + a^4 + a^2 + a^2$. It cannot be simplified and remains as $3a^4 + 2a^2$.

Collect different powers separately. Conventionally we state the larger powers first.

> **EXAMPLE**
>
> $3a^2 + 6a^4 + 2a^2 + a$ can be simplified to
>
> $6a^4 + (3 + 2)a^2 + a$, which further simplifies to
>
> $6a^4 + 5a^2 + a$

10.3 The factors of an expression

In Section 2.1 we learnt that the factors of a number are all the whole numbers that divide into it without a remainder.

Algebraic expressions also have factors.

> **EXAMPLE**
>
> The factors of $30ab^2$ include 2, 3, 5, a and b.
>
> $$30ab^2 = 2 \times 3 \times 5 \times a \times b \times b$$
>
> This can be stated in different ways, for example:
>
> $$30ab^2 = 6(5ab^2)$$
>
> $$30ab^2 = 3b(10ab)$$

10.4 Cancelling in fractions

Cancelling is another way to simplify fractions.

Common factors are those factors that are common to two or more numbers. In order to cancel in algebraic fractions, we find common factors in the numerator and the denominator of the fraction and then we can cancel.

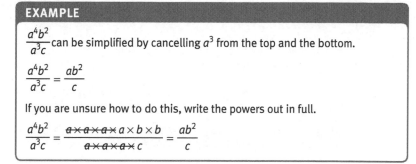

EXAMPLE

$\dfrac{a^4 b^2}{a^3 c}$ can be simplified by cancelling a^3 from the top and the bottom.

$\dfrac{a^4 b^2}{a^3 c} = \dfrac{ab^2}{c}$

If you are unsure how to do this, write the powers out in full.

$\dfrac{a^4 b^2}{a^3 c} = \dfrac{\cancel{a} \times \cancel{a} \times \cancel{a} \times a \times b \times b}{\cancel{a} \times \cancel{a} \times \cancel{a} \times c} = \dfrac{ab^2}{c}$

10.5 Cancelling expressions

We can cancel expressions in the same way that we can cancel single variables.

EXAMPLE

$\dfrac{d^2 (ab + c)^5}{e(ab + c)^3}$ can be simplified by cancelling $(ab + c)^3$ from top and bottom.

$\dfrac{d^2 (ab + c)^5}{e(ab + c)^3} = \dfrac{d^2 (ab + c)^2 \cancel{(ab + c)^3}}{e \cancel{(ab + c)^3}} = \dfrac{d^2 (ab + c)^2}{e}$

10.6 When not to cancel in fractions

If we cannot find a common factor for the whole of the top and the whole of the bottom of the fraction, we cannot cancel in the fraction.

EXAMPLE

$\dfrac{a^4 b^2 + d}{a^3 c}$ cannot be simplified by cancellation, because there is no common factor for $a^4 b^2 + d$ and $a^3 c$.

10.7 Multiplying out expressions

Expressions with brackets can be **multiplied out**.

EXAMPLE

$3(4a + 2b)$ multiplies out to $12a + 6b$.

Make sure that you multiply everything inside the bracket with everything outside.

10.8 The highest common factor

Factorising an algebraic expression is the opposite of multiplying out. To do this, we pull out a common factor, and put the remainder in brackets.

To factorise an algebraic expression, pull out the **highest common factor** – the largest factor that is common to each term.

> **EXAMPLE**
>
> We wish to factorise $12a$ and $6b$.
>
> Separating the components, for 12a and 6b the highest common factor is 6.
>
> Pulling out the 6 and putting the remainder in brackets gives:
>
> $6(2a + b)$.
>
> So, $12a + 6b$ factorised gives $6(2a + b)$.

10.9 The difference of squares

Where there is a **difference of squares**, i.e. when one square is subtracted from another, an expression can be factorised.

> **EXAMPLE**
>
> $a^2 - b^2 = (a - b)(a + b)$
>
> (Note that $(a - b)(a + b) = a^2 - b^2 + ab - ab = a^2 - b^2$)

However, we can't factorise a *sum* of squares.

> **EXAMPLE**
>
> $a^2 + b^2$ cannot be factorised.

Test yourself

The answers are given on pages 210–11.

Question 10.1
Simplify $15a^5 + 12a^3 + 2a^3 + 4a^2 + a^2 + 7a$

Question 10.2
Simplify $\dfrac{a^2b}{a^3} \times \dfrac{a^4b^2}{b^3}$

Question 10.3
Simplify by cancelling:
$$\frac{a^3b^3\,(c+2d)^4}{a^2b^4\,(c+2d)}$$

Question 10.4
Which of these fractions can be simplified by cancelling?

1) $\dfrac{a^2 - b^4}{b}$

2) $\dfrac{c^4d^2 + b^2c^2d^2}{c^2 + b^2}$

3) $\dfrac{e^3d^2 - cf}{cf}$

Question 10.5
Multiply out $5a(2a - b^2)$

Question 10.6
Factorise $6a^3b^2 + 9a^2b^4$

Question 10.7
Factorise $a^2 - 4b^2$

Question 10.8
a) Write the equation that describes volume (V) of a cylinder of length L, given that the area of the circular face is πr^2, where π is a constant and r is the radius of the circle.
b) Given sections of long bone specimens, what measurements would you take in order to make a reasonable approximation of their volume?
c) To calculate a bone's density, we can divide the mass (m) in grams of the bone by its volume (V) in cm^3. Give the equation for the density of sections of long bones in terms of mass, radius and length.
d) If length and radius of the bone specimens are measured in cm and mass is measured in grams, what are the units of bone density?

Question 10.9
The "equations of motion" describe the behaviour of physical systems in terms of their movement as a function of time. Two of the equations are as follows:
$v = u + at$
$v^2 = u^2 + 2as$
where u is the initial velocity, v is the final velocity, a is the acceleration, t is time and s is distance travelled.
Square the first equation and substitute it into the second one, simplify and make s the subject to find the third equation of motion.

Polynomials

In science, some relationships are linear, like the equation $y = mx + c$ for a straight-line graph.

Polynomials describe relationships that contain *powers* of numbers and therefore are not linear.

11.1 The definition of a polynomial

Polynomials are expressions where all the terms have a variable raised to a positive "integer" (i.e. whole number) power greater than one.

EXAMPLE

$5x^4 + 6a^2 - 4x + 3$ is a polynomial.

The **degree** of a polynomial is its highest power.

EXAMPLE

$5x^4 + 6a^2 - 4x + 3$ is a 4th degree polynomial.

11.2 Other names for polynomials

A **binomial** expression is a polynomial with two terms.

A **trinomial** expression is a polynomial with three terms.

A **quadratic** expression is a second degree polynomial: the highest power is 2.

A **cubic** expression is a third degree polynomial: the highest power is 3.

EXAMPLE

$6a^2 - 4x + 3$ has three terms, so is a trinomial. Its highest power is 2, so it is a quadratic expression.

11.3   Adding and subtracting polynomials

In order to add or subtract polynomials, keep each term separate and only add or subtract terms with the same power.

> **EXAMPLE**
>
> When adding $4x^2 + 3x + 6$ to $8x^3 + 2x + 4$, we collect terms which are raised to the same power:
>
> $$\begin{aligned} & 4x^2 + 3x + 6 \\ 8x^3 & + 2x + 4 \\ \hline 8x^3 & + 4x^2 + 5x + 10 \end{aligned}$$
>
> We can use the same process to subtract $x^3 + 6x^2 - 3$ from $4x^2 + 2x + 1$:
>
> $$\begin{aligned} & 4x^2 + 2x + 1 \\ -(x^3) & - (6x^2) - (-3) \\ \hline -x^3 & - 2x^2 + 2x + 4 \end{aligned}$$

11.4   Multiplying polynomials

When multiplying polynomials, every term in the first expression must be multiplied by every term in the second expression.

> **EXAMPLES**
>
> Multiplying the first degree polynomials $x + 2$ and $x + 3$ can be represented by $(x + 2)(x + 3)$.
>
> We need to multiply both x and 2 in the first expression by both x and 3 in the second expression.
>
> $$(x + 2)(x + 3) = x(x + 3) + 2(x + 3)$$
>
> This gives:
>
> x^2, $3x$, $2x$ and 6.
>
> Adding these together gives:
>
> $x^2 + 5x + 6$.
>
> We wish to multiply the terms $x^5 + 3x^4 + 2$ and $6x^2 + 3x - 5$.
>
> Multiplying x^5 with each component of the second term gives $6x^7$, $3x^6$ and $-5x^5$.
>
> Multiplying $3x^4$ with each component of the second term gives $18x^6$, $9x^5$ and $-15x^4$.
>
> Multiplying 2 with each component of the second term gives $12x^2$, $6x$ and -10.

Adding all these together:

$$6x^7 + 3x^6 - 5x^5$$
$$+ 18x^6 + 9x^5 - 15x^4$$
$$+ 12x^2 + 6x - 10$$
$$\overline{6x^7 + 21x^6 + 4x^5 - 15x^4 + 12x^2 + 6x - 10}$$

11.5 Factorising polynomials

The previous chapter explained how to factorise an algebraic expression, by pulling out a common factor and putting the remainder in brackets.

Factorising a polynomial like $4x^3 - 6x^2 + 2x - 3$ means reversing the multiplication and taking it back to $(2x^2 + 1)(2x - 3)$.

Look for the common factors. This can be a bit tricky but it gets easier with practice.

If we represent the constant with the symbol a, many polynomials follow one of the following patterns on the left hand side of the following table. Follow each polynomial across to the right hand side to see how it can be factorised.

Table showing factorisation of common polynomials					
	Polynomial multiplied	↔	Intermediate step	↔	Polynomial factorised
1	$x^2 + 2xa + a^2$	↔	$(x + a)(x + a)$	↔	$(x + a)^2$
2	$x^2 - 2xa + a^2$	↔	$(x - a)(x - a)$	↔	$(x - a)^2$
3	$x^2 - a^2$	↔		↔	$(x + a)(x - a)$
4	$x^3 + 3x^2a + 3xa^2 + a^3$	↔	$(x + a)(x + a)(x + a)$	↔	$(x + a)^3$
5	$x^3 - 3x^2a + 3xa^2 - a^3$	↔	$(x - a)(x - a)(x - a)$	↔	$(x - a)^3$

EXAMPLE

The polynomial
$$x^2 + 8x + 16$$
can be factorised to
$$x^2 + 2(4x) + 4^2$$
When $a = 4$, this is directly equivalent to the first polynomial in the table above:
$$x^2 + 2xa + a^2$$
which we know from the table factorises to
$$(x + a)^2$$
So,
$$x^2 + 8x + 16 \text{ factorises to } (x + 4)^2.$$
You can check this by multiplying out $(x + 4)(x + 4)$.

Test yourself

The answers are given on page 211.

Question 11.1
What is the degree of the following polynomial?
$6a^5 + 5a^3 + 2a^2 - 12$

Question 11.2
Subtract $6x^4 + 9x^3 - x^2 + 5$ from
$2x^5 + 7x^4 + 5x^3 + 4$

Question 11.3
Multiply out $(4x^4 - x^2 + 5)(2x^5 + 3x^2 + 6)$

Question 11.4
Factorise the polynomial $x^2 - 6x + 9$

Algebraic equations

Many relationships in science can be generalised into algebraic equations. Anything from the growth of a child to the photosynthesis rate of a plant can be written as an algebraic equation.

12.1 Balancing the sides of an equation

In an equation, the equals sign can be represented as a balance.

Whatever is done to one side of the equation must also be done to the *whole* of the other side. So, if we add, subtract, multiply or divide something to one side, we need to do the same to the whole of the other side.

> **EXAMPLE**
>
> $$3x + 2 = 11$$
> ▲
>
> Subtracting 2 from each side keeps the equation balanced:
>
> $$3x + 2 - 2 = 11 - 2$$
>
> This can be simplified to:
>
> $$3x = 9$$
>
> Then we can divide both sides by 3:
>
> $$\frac{3x}{3} = \frac{9}{3}$$
>
> So $x = 3$

12.2 Manipulating equations with different powers

We can also manipulate equations with different powers.

> **EXAMPLE**
>
> We wish to solve the equation $ay^2 - b = x$ for y. This means that we wish to manipulate the equation to show what y equals.
>
> We call this "making y the subject of the equation".
>
> Adding b to each side gives
>
> $$ay^2 = x + b$$

Dividing each side by a gives

$$y^2 = \frac{x+b}{a}$$

Taking the square root of each side gives

$$y = \sqrt{\frac{x+b}{a}}$$

Test yourself

The answers are given on page 211.

Question 12.1
Solve $3x^2 = 12$

Question 12.2
Make y the subject of the equation $x = 4y^3 + 1$.

Question 12.3
When you sneeze you accelerate particles in your lungs from rest to 40 m s^{-1} in approximately half a second. Using the equation
$v = u + at$
where v is the final velocity in m s^{-1}, u is the initial velocity in m s^{-1}, a is acceleration in m s^{-2}, t is time in seconds, and
$S = ut + \frac{1}{2}at^2$
where S is the distance travelled in metres:
a) Rearrange the first equation so that acceleration is the subject.
b) Calculate the acceleration of a dust particle on sneezing.
c) To three decimal places, how long will a particle of dust take to travel from an alveolus to the mouth, if the distance is 30 cm?

Question 12.4
a) In respiration,
V_E = the volume of air expired per minute;
V_T = the tidal volume (the volume of air moved in or out of the lungs with each breath);
F = frequency of breaths per minute.
Write an equation relating all three.
b) V_E can also be described in terms of:
V_A, the volume of air that is inspired per minute, reaches the alveoli in the lungs and takes part in gas exchange, and

V_D, the volume of air that is inspired per minute, reaches "dead space" and does not take part in gas exchange.
Write an equation relating V_E, V_A and V_D.
c) V_A is a measurement that can be useful in patient care. Although V_E is easy to measure at the bedside using a spirometer, spirometry cannot differentiate between V_A and V_D. However, CO_2 is only present as a trace gas in inspired air, yet is present in larger concentrations in expired air. Thus, measuring expired CO_2 can be used to calculate V_A, because only the air that reaches the alveoli and takes part in gas exchange will contain CO_2. If V_{CO_2} = the volume of CO_2 excreted by the lungs per minute, and F_ACO_2 = the proportion of air from the alveoli that is CO_2, write an equation for V_{CO_2} in terms of F_ACO_2 and V_A.
d) P_aCO_2 is the "partial pressure" of CO_2 in the arteries, and it is proportional to F_ACO_2. So, $F_ACO_2 = k \times P_aCO_2$, where k is the constant of proportionality.
Rewrite your equation for (c) in terms of P_aCO_2 instead of F_ACO_2.
e) From experiments we know that in this case
$$k = \frac{1}{0.863}$$
Rewrite the equation, this time rearranging it to put V_A as the subject and replacing k with its value.
This equation is useful because the P_aCO_2 and V_{CO_2} can be measured and so V_A can be calculated.

Quadratic equations

Some scientific relationships can be represented by quadratic equations. For example, the equation underlying the Hardy-Weinberg equilibrium in population genetics is a quadratic equation.

13.1 Solving quadratic equations

Quadratic equations have expressions containing a square power, e.g. x^2.

They have two solutions, i.e. x can have two possible values.

Quadratic equations have the form:

$$ax^2 + bx + c = 0$$

where a, b, and c are constants.

EXAMPLE

$x^2 + 2x - 15 = 0$ is an example of a quadratic equation.

See how it relates to the quadratic formula $ax^2 + bx + c = 0$.

In this example, a is 1, b is 2, and c is −15.

x can equal + 3 and it can also equal − 5.

13.2 Different ways to solve quadratic equations

We will describe four ways to solve quadratic equations:

- graphically
- by factorisation
- by using the quadratic formula
- by completing the square

13.3 Graphical solution

If we plot the quadratic equation on a graph, the solutions are the points at which the graph crosses the x-axis.

EXAMPLE

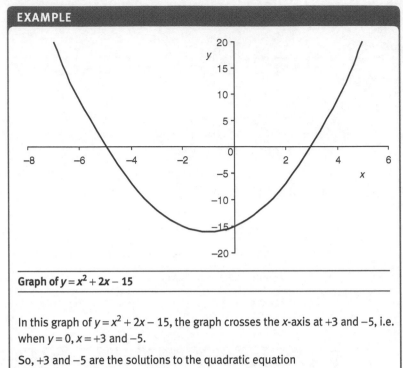

Graph of $y = x^2 + 2x - 15$

In this graph of $y = x^2 + 2x - 15$, the graph crosses the x-axis at +3 and −5, i.e. when $y = 0$, $x = +3$ and −5.

So, +3 and −5 are the solutions to the quadratic equation $x^2 + 2x - 15 = 0$.

13.4 Solution by factorisation

If the quadratic equation is factorised (factorising can be thought of as reversing a multiplication; Chapter 10 explains the basics), the solution is given if each factor is set to zero.

EXAMPLE

$$x^2 + 2x - 15 = 0$$

can be factorised to:

$$(x + 5)(x - 3) = 0$$

If either one of the expressions in brackets is zero, the equation will still equal zero.

For example

$$0(x - 3) = 0 \text{ or } (x + 5)0 = 0$$

To set the first expression, $(x + 5)$, to zero, x must equal −5.

To set the second expression, $(x - 3)$, to zero, x must equal +3.

Thus $x = -5$ or $x = 3$

13.5   Solution by using the quadratic formula

Given the quadratic equation

$$ax^2 + bx + c = 0,$$

the solution can be calculated from the **quadratic formula:**

$$x = \frac{-b \pm \sqrt{b^2 - 4ac}}{2a}$$

Put the numbers into the equation and calculate the two possible values of x.

EXAMPLE

We noted above that for the quadratic equation $x^2 + 2x - 15 = 0$, a is 1, b is 2, and c is -15.

$$x = \frac{-b \pm \sqrt{b^2 - 4ac}}{2a} = \frac{-2 \pm \sqrt{2^2 - (4 \times 1 \times -15)}}{2 \times 1} =$$

$$\frac{-2 \pm \sqrt{4 - (-60)}}{2} = \frac{-2 \pm \sqrt{64}}{2} = \frac{-2 \pm 8}{2}$$

The two results are therefore:

$$\frac{-2 + 8}{2} = 3 \text{ and } \frac{-2 - 8}{2} = -5$$

13.6   Solution by completing the square

This involves creating a square trinomial that we can solve by taking its square root.

EXAMPLE

To solve $3x^2 + 24x - 27 = 0$

Put the x^2 and x terms on one side and the constant on the other:

$$3x^2 + 24x = 27$$

Divide both sides by the "coefficient" (the multiplier) of x^2 (in this case 3):

$$\frac{3x^2}{3} + \frac{24x}{3} = \frac{27}{3}, \text{ therefore}$$

$$x^2 + 8x = 9$$

We take half the coefficient of x, (in this case half of 8, giving 4), square it (giving 16), and add it to both sides.

This gives a square trinomial:

$$x^2 + 8x + 16 = 9 + 16 = 25$$

We can now factorise the left-hand side of the equation:

$$(x + 4)^2 = 25$$

Now we need to take the square root of both sides:

$$\sqrt{(x + 4)^2} = \sqrt{25}$$

As a square root can be positive or negative, there needs to be a $\pm$ on the right side of the equation.

$$x + 4 = \pm 5$$

We can then solve the equation.

$$x = +5 - 4 = 1 \text{ and } x = -5 - 4 = -9$$

Test yourself

The answers are given on page 211.

Question 13.1
Solve by factorisation:
$x^2 + 6x + 8 = 0$

Question 13.2
Solve, using the quadratic formula:
$x^2 + 6x + 8 = 0$

Question 13.3
Solve by completing the square:
$x^2 + 6x + 8 = 0$

Question 13.4
A tracer float is placed in a stream and travels down a mountainside at a vertical velocity of 5 m s^{-1}. The float then passes over a waterfall that is 10 m high. The fall of the float is accelerated by gravity at approximately 10 m s^{-2}.
Using $s = ut + \frac{1}{2}at^2$, where s is distance travelled (m), u is the initial velocity (m s^{-1}), t is time (s), v is the final velocity (m s^{-1}) and a is the acceleration (m s^{-2}), calculate the time taken for the float to reach the bottom of the waterfall.

Simultaneous equations

Simultaneous equations can be used to calculate when two relationships coincide.

In the life and medical sciences, the relationship between two variables may directly reflect on the relationship of two other variables. For example, a change in numbers of an animal species over time may affect the numbers of another species higher up the food chain. We can establish the link by studying the population growth or decay of both simultaneously.

Mathematically, we do this by solving simultaneous equations.

14.1 Two equations with a single solution

Equations that can be solved for values of x and y are called simultaneous equations.

> **EXAMPLE**
>
> $$x - y = 5 \text{ and } x + 2y = -4$$
>
> can both be solved if
>
> $$x = 2 \text{ and } y = -3$$

14.2 Three ways to solve simultaneous equations

We will show three ways to solve simultaneous equations: graphically, by substitution, and by elimination.

14.3 The graphical solution

If we plot the two equations on a graph, the co-ordinates of where they intersect give the solution.

> **EXAMPLE**
>
> To solve the simultaneous equations $x + 3y = 4$ and $6x - 5y = 1$ by graph, we need to plot both lines on a graph. The solution is the point where they intersect.

First, we need to rearrange the two equations to the form $y = mx + c$, so that they can be plotted on the graph.

The equation $x + 3y = 4$ manipulates to $y = -\dfrac{x}{3} + 1.3\dot{3}$.

We found in Chapter 9 that this means that the gradient of the line is $-1/3$, and that the line crosses the y-axis at $+1.3\dot{3}$.

Manipulating the equation $6x - 5y = 1$ gives

$y = 1.2x - 0.2$

so the gradient of this line is 1.2 and it crosses the y-axis at -0.2

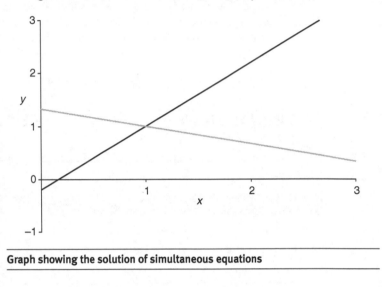

Graph showing the solution of simultaneous equations

The lines intersect each other at (1,1), i.e. where $x = 1$ and $y = 1$, so the solution is $x = 1$, $y = 1$.

14.4 Solution by substitution

For solution by substitution, we manipulate one of the equations so that y is written as an expression of x. This value of x is then substituted into the second equation, and the solution can be calculated.

EXAMPLE

To solve the simultaneous equations $x + 3y = 4$ and $6x - 5y = 1$ by substitution, we manipulate the first equation by subtracting $3y$ from each side:

$x = 4 - 3y$

We now substitute this into the second equation, giving:

$6(4 - 3y) - 5y = 1$

This multiplies out to:

$$24 - 18y - 5y = 1$$

Subtract 24 from each side:

$$-18y - 5y - 1 - 24$$

This simplifies to:

$$-23y = -23$$

Dividing both sides by 23 gives:

$$-y = -1$$

So $y = 1$

Now we know the value of y, we can substitute it into either of the original equations to calculate the value of x:

$$x + 3y = 4$$

$$x + (3 \times 1) = x + 3 = 4$$

$$x = 4 - 3 = 1$$

So the solution is $x = 1$, $y = 1$.

14.5 Solution by elimination

Another way of solving simultaneous equations is by eliminating one of the unknowns (either the x or the y).

EXAMPLE

To solve the simultaneous equations

$$x + 3y = 4 \tag{1}$$

and

$$6x - 5y = 1 \tag{2}$$

by elimination, we can manipulate equation (1) by multiplying both sides by 6:

$$6(x + 3y) = 6 \times 4$$

This multiplies out to:

$$6x + 18y = 24$$

So, the equation (1) can be manipulated to:

$$6x = 24 - 18y$$

By adding $5y$ to both sides of equation (2), we get:

$$6x = 1 + 5y$$

Both sides have now been manipulated to equal $6x$, so both equations are now equal:

$$24 - 18y = 6x = 1 + 5y$$

i.e.:

$$24 - 18y = 1 + 5y$$

This simplifies to:

$$-18y - 5y = 1 - 24$$

and again to:

$$-23y = -23$$

Once more, dividing both sides by 23 gives:

$$-y = -1$$

So $y = 1$

As in the solution by substitution example above, now we know the value of y, we can substitute it into either of the original equations and find that $x = 1$.

Test yourself

The answers are given on pages 211–12.

Question 14.1
Solve $2x + y = 8$ and $3x + 2y = 14$ by substitution.

Question 14.2
Use elimination to solve $2x + y = 8$ and $3x + 2y = 14$.

Question 14.3
There are 1000 people living in a village. Each year the number is expected to increase by 50. The population in another village is 1600, declining by 50 per year.
In how many years will they have an identical population size?
What will the population be at that time?
Solve this in two ways: graphically, and by calculation.

Question 14.4
A study shows that in an area of ocean, light sufficient to support phytoplankton growth penetrates to a depth of 25 m. Nutrient-rich waters sufficient for phytoplankton growth well up to a level of 12 m. Phytoplankton growth occurs where there is sufficient light and nutrients.
a) Between which depths would you expect phytoplankton to grow?
b) We can simplify reality by describing light loss as it penetrates water as being proportional to its depth. In this case $y = -\frac{1}{4}x + 25$, where y is depth (m) and x is the percentage of surface light intensity.
One type of phytoplankton needs at least 75% of the surface light intensity to be able to grow. Would this phytoplankton grow in the column of water?
c) At what minimum percentage of surface light intensity would phytoplankton need to be able to grow in order to survive in this column of water?

Sequences and series of numbers

In nature we can often identify patterns. These patterns can frequently be identified as one of two distinct groups: **arithmetic series** and **geometric series**.

15.1 Sequences

Where we have a **sequence** of numbers, we can use a letter to represent a position in the sequence. Typically, for a simple variable we use the letter n.

So, for the sequence 2, 4, 6, 8, 10 ..., the nth term is $2n$.

To work out the 30th term, we substitute 30 for n, so that $2n = 2 \times 30 = 60$.

15.2 Arithmetic series

In the example above, the difference between each number is constant.

This means that it is an **arithmetic series** (also known as an "arithmetic progression").

An arithmetic series is one in which each number or term is obtained from the previous number or term, by adding (or subtracting) a constant quantity.

This constant is known as the **common difference**.

> **EXAMPLE**
>
> For the 7, 12, 17, 22, 27... sequence, the common difference is +5:
>
> | Term | 1st | | 2nd | | 3rd | | 4th | | 5th |
> |---|---|---|---|---|---|---|---|---|---|
> | Number | 7 | | 12 | | 17 | | 22 | | 27 |
> | Common difference | | +5 | | +5 | | +5 | | +5 | |

An arithmetic series can be symbolised by:

$a, (a + d), (a + 2d), (a + 3d) \ldots \ldots$ to n terms,

where a is the first term (7 in the example above), d is the common difference (5 in the example above), and n symbolises the "term" number (the number that we want to count to).

The nth term for any arithmetic series is given by:

$$a_n = a + (n - 1)d$$

EXAMPLE

For the sequence 13, 16, 19, 22, 25 ..., we wish to calculate the 78th term.

The first term is 13, so $a = 13$; the common difference is 3, so $d = 3$; we want the 78th term, so $n = 78$.

Substituting these into the equation

$$a_n = a + (n - 1)d$$

we get

$$a_{78} = 13 + (78 - 1)3 = 244$$

So the 78th term is 244.

15.3 The sum of an arithmetic series

We use the Σ ("summation") symbol to indicate that we are adding a list of terms together. We then put a comma between each term, rather than a plus sign.

Adding the arithmetic series

$$a + (a + d) + (a + 2d) + (a + 3d) \dots \text{ to } n \text{ terms}$$

can therefore be symbolised by:

$$\sum [a, (a + d), (a + 2d), (a + 3d) \dots (a + (n - 1)d)]$$

The formula that calculates this total is:

$$S_n = \frac{n}{2}[2a + (n - 1)d]$$

where S_n is the sum of the first n terms.

While this may look complex, note that it is the formula for the nth term for any arithmetic series

$$a + (n - 1)d$$

plus the first number in the series, a, the total being multiplied by $\frac{1}{2}n$.

Try not to confuse the S_n notation (meaning the sum of the first n terms) with the log base notation, $\log_n$.

EXAMPLE

Given the sequence 13, 16, 19, 22, 25 ..., we wish to calculate the sum of the first 78 terms.

The first term is 13, so $a = 13$; the common difference is 3, so $d = 3$; $n = 78$.

Substituting these into the formula

$$S_n = \frac{n}{2}[2a + (n-1)d]$$

gives

$$S_{78} = \frac{78}{2}[(2 \times 13) + (78 - 1)3] = 39(26 + 231) = 10\,023$$

So the sum of the first 78 terms is 10 023.

15.4 Geometric series

A **geometric series**, also known as a "geometric progression", is formed by multiplying each term by a constant. This constant is known as the **common ratio** and can be any value except 0, 1, or −1.

EXAMPLE

For the 3, 6, 12, 24, 48 ... sequence, each number is multiplied by 2:

Term	1st		2nd		3rd		4th		5th
Number	3		6		12		24		48
Common ratio		×2		×2		×2		×2	

A geometric series can be symbolised by:

a, ar, ar^2, ar^3 ... to n terms,

where:

a is the first term (3 in the example above),

r is the common ratio (2 in the example above), and

n is the number of terms.

The nth term for any geometric series is given by:

$$a_n = a\,r^{n-1}$$

EXAMPLE

For the geometric series 5, 20, 80, 320, 1280 ..., we wish to calculate the 9th term.

Here, the first term is 5, so $a = 5$. The common ratio is 4, so $r = 4$. The nth term is the 9th, so $n = 9$.

Substituting these into the formula

$$a_n = a\,r^{n-1}$$

we get

$$a_n = 5 \times 4^{9-1} = 5 \times 4^8 = 5 \times 65\,536 = 327\,680$$

15.5 The sum of a geometric series

The sum of any geometric series can be written as

$$S_n = \frac{a(r^n - 1)}{r - 1}$$

You may also see it written as

$$S_n = \frac{a(1 - r^n)}{1 - r}$$

which gives the same answer, but is easier to use if $r < 1$.

EXAMPLE

In calculating the sum of the first nine terms of the series 5, 20, 80, 320, 1280 ..., again $a = 5$, $r = 4$, and $n = 9$.

Substituting these into the formula above gives:

$$S_9 = \frac{5(4^9 - 1)}{4 - 1} = \frac{5 \times 262\,143}{3} = 436\,905$$

So the sum of the terms is 436 905.

Test yourself

The answers are given on page 213.

Question 15.1
There are initially 12 robins in an area of parkland. The yearly count of robins increases as follows: 16, 20, 24, 28. Assuming that the rate of increase remains constant, how many robins would be expected in the 10th year?

Question 15.2
If the life expectancy of the robins in the previous question is 1 year, how many robins in total would have inhabited the parkland in the 10 years?

Question 15.3
Halley's comet becomes visible to the naked eye every 76 years. It was last visible in 1986.
a) Is this an arithmetic or geometric progression?
b) What is the equation for the nth sighting, if the 1986 sighting is $n = 1$?
c) Use the equation to find out when the next two sightings will be. (Hint: when $n = 2$ and $n = 3$.)

Question 15.4
Domestic bees make their honeycomb by starting with a single hexagonal cell, then forming a ring of hexagonal cells around the first cell, then a second ring around the first and so on.
a) Is this an arithmetic or geometric progression?
b) What is the equation for the number of cells in the nth ring?
c) What is the equation that gives the total number of cells present at the completion of each ring?

d) Use the last equation to calculate the total cells present at the completion of the 14th hexagon. (Hint: remember to include the first cell.)

Question 15.5
One year after a forest fire, in one quadrat (measured area) there are 250 plants which are greater than 100mm high. In the succeeding years the numbers go up to 750, 2250 and 6750. Assuming the numbers increase in the same pattern each year, how many plants would we expect to find 7 years after the forest fire?

Question 15.6
If the plants studied in the previous question were all annuals, what is the total number of plants that would have grown in the plot during the 7 years?

Question 15.7
The population of India in 2011 was 1.241 billion people, to 4 significant figures. The growth rate is 1.4% per year.
a) Is this an arithmetic or geometric progression?
b) Take the first term ($n = 1$) as the population in 2011, $n = 2$ as the population in 2012, etc. and assume that the growth rate remains constant. Use these values to write the equation for the population of India.
c) What would we expect India's population to be in 2013? Leave the numbers in billions.

Working with powers

This chapter describes a number of rules that can help us when working with powers.

16.1 The power of zero

In Section 5.2, we found that $10^0 = 1$.

This applies for any value (except zero itself). Any value to the power of zero is 1.

EXAMPLE

$x^0 = 1$

16.2 Useful rules for working with powers

Other useful rules for working with powers are as follows:

$$x^{-2} = \frac{1}{x^2}$$

$$x^{\frac{1}{2}} = \sqrt{x}$$

$$x^{\frac{1}{3}} = \sqrt[3]{x}$$

$$x^{\frac{2}{3}} = \sqrt[3]{x^2} = \left(\sqrt[3]{x}\right)^2$$

$$x^2 \times x^3 = x^{2+3} = x^5$$

$$\frac{x^5}{x^3} = x^{5-3} = x^2$$

$$\left(x^2\right)^3 = x^6$$

$$(xab)^2 = x^2 a^2 b^2$$

$$\left(\frac{x}{a}\right)^2 = \frac{x^2}{a^2}$$

16.3 Adding and subtracting powers

As seen in Section 5.5, numbers in standard form can be added or subtracted if the powers are the same. The same applies to algebraic expressions.

> **EXAMPLE**
>
> $$5x^3 - 2x^3 = 3x^3$$
>
> However,
>
> $$x^2 + x^5$$
>
> cannot be added, as the powers are different.

16.4 Working with roots

All roots can be converted into powers.

> **EXAMPLES**
>
> $$\sqrt[3]{x} = x^{\frac{1}{3}} = x^{0.33}$$
>
> $$\sqrt[4]{x^2} = x^{\frac{2}{4}} = x^{\frac{1}{2}} = x^{0.5}$$

In basic mathematics, we can't have a square root (or any even-numbered root) of a negative value.

$$\cancel{\sqrt{-x}}$$

$$\cancel{\sqrt[4]{-x}}$$

You may find these root rules helpful:

$$\sqrt[3]{x} \times \sqrt[3]{a} = \sqrt[3]{xa}$$

$$\frac{\sqrt[3]{x}}{\sqrt[3]{a}} = \sqrt[3]{\frac{x}{a}}$$

$$\sqrt[2]{\sqrt[3]{x}} = \sqrt[2\times3]{x} = \sqrt[6]{x}$$

Test yourself

The answers are given on page 213.

Question 16.1

Calculate $\dfrac{(a+2)^7}{(a+2)^5}$

Question 16.2

Calculate $\sqrt[3]{(2a-1)^6}$

Question 16.3

The Body Mass Index (BMI) $= \dfrac{\text{mass (kg)}}{[\text{height (m)}]^2}$

A Healthcare Assistant noted that a patient had a BMI of 38 and a mass of 124 kg, however, he forgot to write down the patient's height.

Calculate the patient's height, giving your answer to three significant figures.

Question 16.4
Poiseuille's law can be used to calculate the flow (Q) of liquid along a tube:

$$Q = \frac{\pi P r^4}{8 \eta L}$$

where P = pressure, r = radius of tube, η = coefficient of viscosity and L = length of tube. What effect does doubling the diameter have on the flow rate?

Question 16.5
The speed of light is 3.0×10^8 m s^{-1} and the speed of sound is 3.4×10^2 m s^{-1}, both to two significant figures.
A bolt of lightning produces light and sound simultaneously. How long does it take for the sound to travel the same distance that the light travels in one ten thousandth of a second? Give your answer to the nearest second.

Question 16.6
Bytes are used to measure storage capacity in computers; 1 kilobyte = 2^{10} bytes = 1024 bytes (Note: not 1000).
So 1 kilobyte = 2^{10} bytes
1 megabyte = $2^{10} \times 2^{10} = 2^{20}$ bytes
1 gigabyte = $2^{10} \times 2^{10} \times 2^{10} = 2^{30}$ bytes
A tablet computer is sold with 64 GB (gigabytes) of storage capacity. An old 3.5" floppy disk held 1.44 MB (megabytes) of storage capacity. How much more capacity has the 64 GB tablet than the floppy disk? Give your answer to two significant figures.

Logarithms

Many biological and biochemical systems are logarithmic. For example, population growth can show logarithmic properties; the pH measure of acidity is a logarithmic measure.

17.1 Introducing logarithms

We have already met powers, like 2^3.

The first number, 2 in this case, is called the "base".

The power, 3 in this case, is called the "exponent".

The exponent is the **logarithm** ("log") of the base.

> **EXAMPLES**
>
> $2^3 = 8$ is saying the same as $\log_2 8 = 3$ (spoken as "log base 2 of eight equals 3").
>
> With 10^5, the exponent is 5, therefore the log of 10^5 is 5. This the same as stating that $\log_{10}$ of $100\,000$ is 5, i.e. $\log_{10} 100\,000 = 5$.

Conventionally, if the base is 10 we don't write the 10.

> **EXAMPLE**
>
> $\log_{10} 100 = 2$ is written as $\log 100 = 2$ (spoken as "log 100 equals 2").

Scientific calculators and computer software will calculate logarithms for you: use the "log" key or function.

To convert a logarithm back to its original number, use the "inverse" (or "shift") key and then the log key.

17.2 The natural logarithm

Log base e (e is a constant ≈ 2.718) is called the **natural logarithm**. It is conventionally written ln rather than $\log_e$.

e has the property that $\ln e^x = x$ (i.e. $\log_e e^x = x$).

So, $\ln e = 1$ (remember that ln e is the same as $\ln e^1$, and $\ln e^1 = 1$).

This is important because, if we have an **exponential** relationship, a graph of ln y against x will give a straight line.

Chapter 18 explains the use of e, natural logarithms and exponential relationships in more detail.

17.3 Logarithm rules

The following rules may prove useful:

$\log_a 1 = 0$ (this is the same as saying $a^0 = 1$)

$\log_a a = 1$ (this is the same as $a^1 = a$)

$\log_a (bc) = \log_a b + \log_a c$, i.e. to multiply two numbers, we *add* their logarithms. Remember that logarithms are indices, and we add indices when multiplying.

$\log_a (b/c) = \log_a b - \log_a c$, so to divide one number into another, in the same way that we subtract indices, we *subtract* their logarithms.

$\log_a b^c = c \log_a b$

$\log_a a^c = c$ (because $\log_a a^c = c \log_a a$, and $\log_a a = 1$)

EXAMPLES

ln 1 = 0 (i.e. $\log_e 1 = 0$)

$\log(1546 \times 4326) = \log 1546 + \log 4326 \approx 3.189 + 3.636 = 6.825$

$\log_2 56^4 = 4 \log_2 56$

Test yourself

The answers are given on page 213.

Question 17.1
We will cover pH in detail in Chapter 27. It is a logarithmic scale of hydrogen ion concentration and is defined by:
$pH = -\log[H^+]$
where $[H^+]$ symbolises the concentration of hydrogen ions.
1) If $[H^+] = 1.2 \times 10^{-5}$, what is the pH?
2) If the pH is 6.3, what is $[H^+]$?
Use the "log" and "inverse" (or "shift") keys of a calculator to help you.

Question 17.2
What is ln e^4, where e is the constant ≈ 2.718?
Hint: you can do this without using your calculator.

Question 17.3
The number of bacteria in a culture of *E. coli* has increased from 10 000 to 10 000 000 in 3 hours. In a controlled nutrient base the *E. coli* grow by binary fission (i.e. they reproduce by splitting into two).
The equation to calculate the number of bacteria at a given time interval is:
$b = B \times 2^n$
where b = number of bacteria at the end of the time interval, B = number of bacteria at the beginning of the time interval and n = number of generations (the number of times the cell population doubles during the time interval).

The time taken for a bacterial population to double is called the "generation time", G:

$$G = \frac{t}{n}$$

where t = time interval in hours.
Calculate the generation time in minutes for this culture. Give your answer to two significant figures.

(Hint: use $b = B \times 2^n$ to find n. You will need a calculator for this. Then substitute into the second equation to find the generation time.)

Exponential growth and decay

Many life science systems have an **exponential** relationship with time.

For example:

- the number of bacteria in a culture may double every hour, an example of **exponential growth**;
- the radioactivity of technetium-99m halves every 6 hours, an example of **exponential decay**.

18.1 Formulae for exponential growth

A simple formula that describes exponential growth is

$$y = a^x$$

where a is a constant that depends on the system being studied, and x is the exponent (also called the power, or the index).

Taking the logarithm of both sides of the general equation gives

$$\log y = x \log a$$

So, if we plot $\log y$ against x, an exponential relationship will plot as a straight line with a gradient of $\log a$.

In this example, we have used log to the base 10. However, we will still get a straight line if the log to any base is used.

A relationship that is exponential will give a straight-line graph if plotted on semi-log graph paper (graph paper that has a logarithmic scale on one axis). There is an example of this in Section 24.4.

A more general formula for exponential growth can be written in the form

$$y = ae^{bx}$$

Where a and b are constants that depend on the system.

Taking the natural logarithm of both sides of this equation gives:

$$\ln y = \ln(ae^{bx})$$

Using logarithm rules for multiplication, this is the same as

$$\ln y = \ln a + bx \ln e$$

However, we know that ln e = 1, so

$$\ln y = bx + \ln a$$

From this we can see that by plotting ln y against x, we will get a straight-line graph with a gradient of b and a y-axis intercept of ln a.

18.2 The growth–decay formula

You may see the formula for exponential growth and decay, $y = ae^{bx}$, stated as:

$$N = N_0 e^{kt}$$

where N is the exponentially changing quantity, t is time, N_0 is its value at time $t = 0$, k is the growth constant or decay constant, and e is the constant ≈ 2.718.

EXAMPLE

100 bacteria ($N_0 = 100$) are plated out onto agar ($t = 0$ hours). Five hours later ($t = 5$ hours), there are 300 bacteria ($N_5 = 300$). If we assume exponential growth, we can calculate the growth constant.

$$300 = 100e^{5k}$$

This can be manipulated to:

$$\frac{300}{100} = 3 = e^{5k}$$

Now we take the natural logarithm of both sides of the equation. Remember that the natural logarithm of e^x is x, so ln $e^{5k} = 5k$

$$\ln 3 = 5k$$

$$\frac{\ln 3}{5} = k$$

$$k = 0.22 \text{ per hour}$$

18.3 The doubling time

We can also use the **growth–decay** formula to calculate the **growth constant** if we know how often a population doubles.

EXAMPLE

Moore's law suggests that the power of the fastest computer chips doubles every 18 months.

We could use any initial value for this, say $N_0 = 1$. After 18 months ($t = 18$ months), we have double the initial value, $N_{18} = 2$.

Substituting these values into $N = N_0 e^{kt}$, we get

$$2 = 1e^{18k}$$

This can be manipulated to:

$$\frac{2}{1} = 2 = e^{10k}$$

Taking the natural logarithm of each side gives:

$$\ln 2 = 18k$$

$$\frac{\ln 2}{18} = k$$

$$k = 0.039 \text{ per month}$$

18.4 Exponential decay

We also use the growth–decay formula for exponential decay.

EXAMPLE

The radioisotope technetium-99*m* is a short-lived isotope used in nuclear medicine in the diagnosis of various disorders. It has a half-life of 6 hours.

Again, we can use any initial value for this, say $N_0 = 1$. After 6 hours ($t = 6$ hours), we have *half* the initial value, $N_6 = 0.5$.

Substituting these values into $N = N_0 e^{kt}$, we get

$$0.5 = 1e^{6k}$$

which is then manipulated to:

$$\frac{0.5}{1} = 0.5 = e^{6k}$$

$$\ln 0.5 = 6k$$

$$\frac{\ln 0.5}{6} = k$$

$$k = -0.116 \text{ per hour.}$$

Note that when there is exponential decay, the constant is negative.

18.5 Using the growth constant

Given the growth constant and an initial population size, we can use the growth–decay formula to calculate the population at any given time.

EXAMPLE

Using the bacterial incubation growth constant of $k = 0.22$ per hour given above, we can calculate how many bacteria would be present after 24 hours ($t = 24$ hours), given an initial ($t = 0$ hours) inoculation of 5000 bacteria ($N_0 = 5000$).

$N_{24} = N_0 e^{kt} = 5000e^{0.22 \times 24} = 5000e^{5.28} = 5000 \times 196 =$ 980 000 bacteria

---

Test yourself

The answers are given on pages 213–14.

Question 18.1
A highly infectious virus had affected five patients when first diagnosed. Three weeks later there are 25 patients. Assuming exponential growth of the outbreak, what is the growth constant? Use your calculator to calculate the natural logarithm.

Question 18.2
If the outbreak continues to spread at the same rate, use your calculated growth constant to predict how many patients will be affected in another 4 weeks. Again, use your calculator.

Question 18.3
The doubling time of a culture of E. coli is 20 minutes. Use the growth–decay formula to calculate the growth constant. (Hint: $N = 2N_0$ and $t = 20$ minutes.)

Question 18.4
In a synthetic medium *Mycobacterium tuberculosis* has a growth constant of 7.7×10^{-4} per minute. Using the growth–decay formula, how many bacteria would be present after 72 hours, having started with a single Mycobacterium? (Hint: take care with the units.)

Question 18.5
The radioisotope iodine-131 has a half-life of 8 days. Calculate the decay constant.

Question 18.6
Thallium 201 is a radioactive isotope used for cardiac investigations. What is its half-life, given that the decay constant is -9.5×10^{-3} per hour?

Differential calculus

In this book we cover two types of calculus:

- differential calculus
- integral calculus

Also known as "differentiation", **differential calculus** is concerned with rates of change. It calculates the gradient of a graph at a particular point.

19.1 Constant speed …

The speed of a car is the rate at which the distance the car travels changes with time, for instance miles per hour, or metres per second. This can be calculated by taking the gradient from a distance/time graph.

This is a graph for a car going at a constant speed:

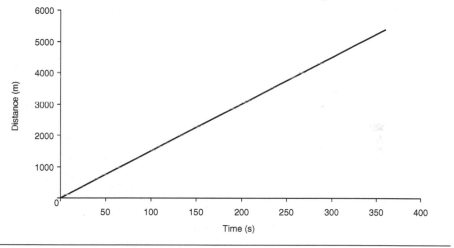

Graph of a car going at a constant speed

The speed of the car in metres per second $= \dfrac{\text{Distance travelled (m)}}{\text{Time taken (s)}}$

In Chapter 9, we learnt that the equation for a straight-line graph is $y = mx + c$, where m is the gradient, and c the point where the line crosses the y-axis.

In this example, the car travels 5400 metres in 360 seconds at a constant speed. The gradient m is therefore $\dfrac{5400}{360} = 15\,\mathrm{m\,s^{-1}}$.

The intercept on the y-axis is 0, i.e. at the start of the journey the car hasn't moved, so the constant c = zero.

The equation for the line is therefore:

$$y = 15x + 0, \text{ or } y = 15x$$

where y is distance in metres, and x is time in seconds.

We also learnt that the gradient of a straight-line graph can be described as:

$$\frac{\text{Change in } y}{\text{Change in } x} = \frac{y_2 - y_1}{x_2 - x_1}$$

This is the rate of change of y with respect to x.

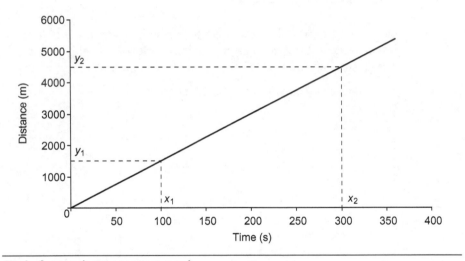

Graph of a car going at a constant speed

In this example, $\dfrac{y_2 - y_1}{x_2 - x_1} = \dfrac{4500 - 1500}{300 - 100} = \dfrac{3000}{200} = 15\,\mathrm{m\,s^{-1}}$.

In calculus the gradient, $\dfrac{y_2 - y_1}{x_2 - x_1}$, is called the **derivative**.

So, the derivative $\dfrac{y_2 - y_1}{x_2 - x_1} = 15\,\mathrm{m\,s^{-1}}$. The graph is a straight line, so the car's speed, and the derivative $\dfrac{y_2 - y_1}{x_2 - x_1}$, is constant throughout the journey.

19.2 ... or acceleration

In the next graph, the car is accelerating, so the speed is increasing throughout the time-span shown.

To calculate the gradient of a curve isn't as easy as with a straight line – the car is constantly accelerating, so the car's speed is constantly changing with time.

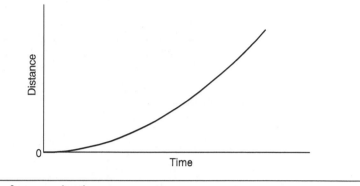

Graph of a car accelerating

19.3 Calculating the gradient of a curve

The formula for the next curve is $y = \frac{1}{2}x^2$.

The gradient (or speed) is constantly changing, so it is different at each point of the graph.

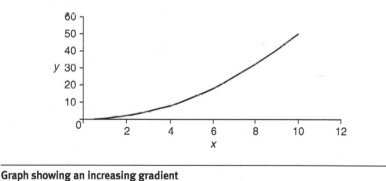

Graph showing an increasing gradient

At any specific point we can calculate the gradient by drawing the **tangent** (a straight line touching the curve at that point) and calculating the gradient of this straight line.

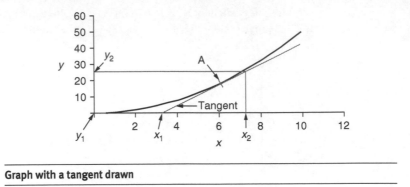

Graph with a tangent drawn

At point A, the gradient of the tangent, and therefore of the curve, is:

$$\frac{y_2 - y_1}{x_2 - x_1} = \frac{25 - 0}{7.5 - 3.5} = \frac{25}{4} = 6.25$$

Differential calculus, otherwise known as **differentiation**, is the mathematical approach to find this gradient (differential).

19.4 Functions

A **function** is a relationship between two or more things, where the value of one thing depends on the value of the other.

> **EXAMPLES**
>
> - The distance that a fish has swum depends on the time it has been swimming.
> - The number of bacteria in a sample depends on the size of the sample.
> - In the graph $y = \frac{1}{2}x^2$ the value of y depends on half the square of x.

In each of these relationships, there is a **dependent variable** (distance, number of bacteria, and y in these examples) and an **independent variable** (time, sample size, x).

Usually, we want to know what the value of the dependent variable is for a given independent variable. So, for $y = \frac{1}{2}x^2$ we might want to know the value of y for a given value of x.

19.5 Function notation

In an equation, one way to write the function $y = \frac{1}{2}x^2$ is to write $f(x) = \frac{1}{2}x^2$.

$f(x)$ is spoken as "f of x".

Both these functions state exactly the same thing.

$f(x)$ doesn't mean f times x. $f(x)$ is simply another way of writing y, i.e. it states that y is a function of x, for example the distance a fish has swum is a function of the time spent swimming.

19.6 Problems with curves

We stated that to be able to calculate a gradient at a particular point, we need to know the change in y for a given change in x, i.e.
$$\frac{(y_2 - y_1)}{(x_2 - x_1)}.$$

To do this, we need to take a section of the graph that has a straight line – and the problem with a curve is that it has no straight line.

For a curve, we can get an *approximation* of the gradient at a particular point by taking a line that intersects the curve at two points (a **secant line**).

EXAMPLE

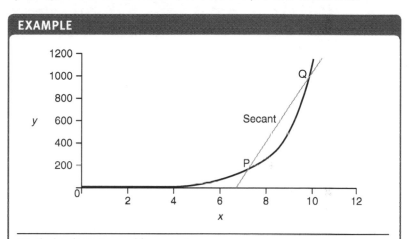

Graph showing a curve with a secant

Here, to get an approximation of the gradient at point P, we have drawn a secant line that intersects the curve at point P and point Q.

$$\text{Secant line gradient} = \frac{y_2 - y_1}{x_2 - x_1} = \frac{1000 - 200}{9.75 - 7.5} = \frac{800}{2.25} = 355.\dot{5}$$

However, if we add a tangent at point P, as in the following graph, we can see that the secant line gradient is steeper than the tangent line.

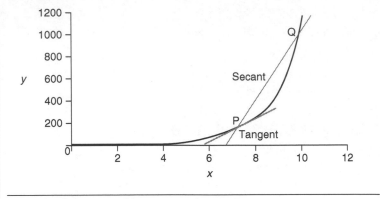

Graph showing a curve with a secant and a tangent

We can use a shorter secant line, from point P to point R:

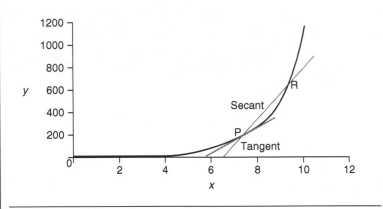

Graph showing a curve with a short secant

Now the secant line gradient $= \dfrac{y_2 - y_1}{x_2 - x_1} = \dfrac{650 - 200}{9 - 7.5} = \dfrac{450}{1.5} = 300.$

This is closer to the tangent's gradient, but still only an approximation of it.

You will see that the shorter the secant, the closer the secant's gradient gets to the gradient of the curve at P.

Now comes one of the key concepts underlying calculus – when the gradient of the secant reaches its **limit**, i.e. when the secant is infinitesimally small, it is, effectively, the gradient of the curve at that point. The notation for this is

$$\lim_{\delta x \to 0} \frac{y_2 - y_1}{x_2 - x_1}$$

In other words, as the difference in x (i.e. $x_2 - x_1$, symbolised here by δx) approaches its limit of zero (symbolised by $\lim_{\delta x \to 0}$), we reach the gradient of the tangent at P.

19.7 The gradient of the chord

With the following curve, the position of point A is given by its co-ordinates *x, y*.

$y_2 - y_1$ is the difference between y_2 and y_1 and it can also be written as δy (which stands for "change in y"). It is pronounced "delta y".

So, the position of point B is given by its co-ordinates $(x + \delta x, y + \delta y)$.

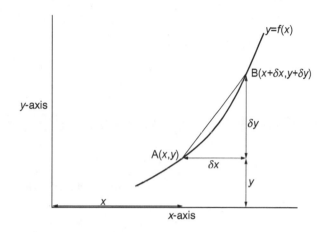

Graph showing the gradient of a chord

The gradient of the chord approaches the tangent of point A as we move point B closer and closer to point A.

At the same time, δy and δx approach their limit of zero.

So, at point A the gradient of the chord AB is:

$$\frac{(y + \delta y) - y}{(x + \delta x) - x}$$

which changes to the mathematical expression $\dfrac{dy}{dx}$ as δx approaches its limit of zero.

$\dfrac{dy}{dx}$ is known as the **differential coefficient**. It is the **derivative** of y with respect to *x*.

Differentiation is the process of calculating this derivative.

Note that $\dfrac{dy}{dx}$ isn't a fraction and the *d* and *y* cannot be separated: the *d* implies a limit.

Putting these together,

$$\lim_{\delta x \to 0} \frac{y_2 - y_1}{x_2 - x_1} = \frac{dy}{dx}$$

$\dfrac{dy}{dx}$ is known as the differential coefficient, the derivative of y, or y'.

As y is a function of x (i.e. $y = f(x)$) the derivative may also be written as $f'(x)$.

Because y is a function of x, the co-ordinates A and B can be written completely in terms of x, as follows:

the co-ordinates of A are (x, y).

So, as y is a function of x (i.e. $y = f(x)$), the co-ordinates of A can also be written $(x, f(x))$.

The co-ordinates of B can therefore be written:

$$(x + \delta x, f(x + \delta x))$$

As the chord between A and B gets smaller, δx gets smaller and moves towards its limit of zero, and A gets closer to B.

19.8 Calculating the differential of x^2

One common function is $y = x^2$.

We want to know the differential (differential coefficient) dy/dx for this function.

If $\qquad\qquad\qquad\qquad\qquad y = x^2$

then $\qquad\qquad\qquad\qquad y + \delta y = (x + \delta x)^2$

Multiplying out $(x + \delta x)^2$ gives:

$$y + \delta y = x^2 + 2x\,\delta x + \delta x^2$$

Substituting in $y = x^2$ then subtracting x^2 from each side gives:

$$\delta y = 2x\,\delta x + \delta x^2$$

Dividing both sides by δx gives the differential coefficient:

$$\frac{dy}{dx} = 2x + \delta x$$

Now here's the calculus equivalent of sleight of hand – we're interested in an infinitesimally small value for δx, i.e. when it approaches zero.

When δx approaches zero,

$$\frac{dy}{dx} = 2x + 0 = 2x$$

So, for the equation $y = x^2$ we've calculated that the differential is $2x$.

19.9 Differentiating with a constant

In the equation for a graph, the constant has no effect on the gradient.

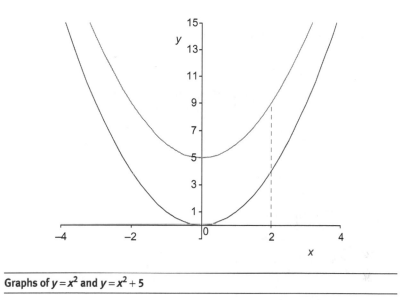

Graphs of $y = x^2$ and $y = x^2 + 5$

In these graphs of $y = x^2$ and $y = x^2 + 5$, for any given value of x the gradient is the same. For instance, where the line $x = 2$ crosses the graphs, the gradient is 4 for both graphs.

The differential of the constant is zero, so the value of the constant (how far up the y-axis the graph starts) doesn't affect, or appear in, the differential.

So, the differentials of $y = x^2$ and $y = x^2 + 5$ are both $\dfrac{dy}{dx} = 2x$.

19.10 Using the differential of x^2

The formula for the area of a circle is $A = \pi r^2$, where A is the surface area, π is a constant 3.1416 and r is the radius of the circle. We can differentiate this formula to calculate the rate of increase of the area at any given radius.

> **EXAMPLE**
>
> In the previous section we have shown that when $y = x^2$, the differential is $2x$, so:
>
> $$\frac{dA}{dr} = \pi(2r) \approx 3.1416(2r) = 6.2832r$$
>
> At the point that the radius of a circle enlarges to 10 mm, the area of the circle is 314.16 mm^2.

At that point, the rate of increase of area is:

$$\frac{dA}{dr} = 6.2832 \times 10 = 62.832 \text{ mm}^2 \text{ per mm.}$$

19.11 The differential of x^3

For the equation $y = x^3$, $\frac{dy}{dx} = 3x^2$.

EXAMPLE

The formula for the volume of a sphere is

$$V = \frac{4}{3}\pi r^3$$

where V is the volume, π is the constant 3.1416 and r is the radius of the sphere.

For the equation

$$y = x^3$$

the differential is

$$\frac{dy}{dx} = 3x^2$$

so for a sphere

$$\frac{dV}{dr} = \frac{4}{3}\pi(3r^2) = 4\pi r^2 \approx 12.57\, r^2$$

When the radius of the sphere enlarges to 10 mm, the volume is increasing by:

$$\frac{dV}{dr} = 12.57(10^2) = 1257 \text{ mm}^3 \text{ per mm.}$$

19.12 The differential of x^n

We've stated that the differential of

$$x^2 \text{ is } 2x$$

and that the differential of

$$x^3 \text{ is } 3x^2.$$

The differential of

$$x^4 \text{ is } 4x^3$$

The differential of

$$2x^4 \text{ is } 4 \times 2x^3 = 8x^3$$

You may have spotted the pattern.

The equation to differentiate x to any power, i.e. $y = mx^n$ is as follows:

$$\frac{dy}{dx} = nmx^{n-1}$$

The constant c in an equation is effectively cx^0 (i.e. $c \times 1$), and the differential of

$$cx^0 \text{ is } 0cx^{0-1} = 0$$

so the differential of a constant is always zero.

EXAMPLE

The differential of $y = 3x^{15} + 7$ is

$$\frac{dy}{dx} = 15(3x^{15-1}) + 0 = 45x^{14}$$

19.13 The differential of $y = \dfrac{1}{x}$

$y = \dfrac{1}{x}$ is the same as writing $y = x^{-1}$.

We can use the $\dfrac{dy}{dx} = nmx^{n-1}$ formula for this.

$$\frac{dy}{dx} = -1x^{-1-1} = -x^{-2}$$

19.14 The differential of e^x

The differential of e^x is e^x, i.e. for e^x, $\dfrac{dy}{dx} = e^x$.

It is the only function (apart from zero) which is equal to its own derivative.

The number that satisfies this is approximately 2.718, so e $\approx$ 2.718

You may recognise the e $\approx$ 2.718 constant from the chapters on logarithms and exponential growth, Chapters 17 and 18.

Test yourself

The answers are given on page 214.

Question 19.1
The heat loss of an organism depends on the surface area of that organism. At a given ambient temperature, the heat loss in a species is found to be $y = 50x^2$, where y is the heat loss in watts and x is the length of the animal in metres.
What is the rate of increase of heat loss when the animal is growing through 1.2m in length?

Question 19.2
A number of wasps in a colony increases with the cube of its radius, described by the formula
$$y = \frac{x^3}{60}$$
where y = number of wasps and x = radius in mm.
What is the rate of increase in number of wasps when the colony has reached 30mm in radius?

Question 19.3
What is the differential of $y = 3x^{20} - 8$ with respect to x?

Question 19.4
What is the differential of $y = \dfrac{3}{x^5}$ with respect to x?

Further differential calculus

This chapter explains how to differentiate more complicated functions.

20.1 The chain rule

The **chain rule** equation is: $\dfrac{dy}{dx} = \dfrac{dy}{dt} \cdot \dfrac{dt}{dx}$.

The chain rule is a technique for calculating the derivative of a function that has a second function applied to it.

An example is the equation $y = (2x-1)^3$. In this case $(2x-1)$ is a function and cubing it applies a second function to it.

Using function notation these can be written as $f(x) = (2x-1)$ and $y = f(x)^3$.

Expanding the equation gives $y = (2x-1)(2x-1)(2x-1) = 8x^3 - 12x^2 + 6x - 1$.

Calculating these expansions can be a lengthy process and the chain rule is used as a shortcut.

How it works
Let y be the function of t and t be a function of x. A small change, δx, in the variable x leads to small changes in y and t, δy and δt respectively.

$$\frac{\delta y}{\delta x} = \frac{\delta y}{\delta t} \cdot \frac{\delta t}{\delta x}$$

As δx, δy and δt tend to zero, this gives the equation for the chain rule:

$$\frac{dy}{dx} = \frac{dy}{dt} \cdot \frac{dt}{dx}.$$

Using the chain rule
To use chain rule first define t, then calculate $\dfrac{dt}{dx}$.

Then write y in terms of t, then calculate $\dfrac{dy}{dt}$.

Then use $\dfrac{dt}{dx}$ and $\dfrac{dy}{dt}$ in the chain rule equation to calculate $\dfrac{dy}{dx}$.

EXAMPLES

Find $\dfrac{dy}{dx}$, where $y = (2x - 1)^3$.

Let $t = 2x - 1$, so $\dfrac{dt}{dx} = 2$.

$$y = t^3, \text{ so } \frac{dy}{dt} = 3t^2$$

$$\frac{dy}{dx} = \frac{dy}{dt} \cdot \frac{dt}{dx} = 3t^2 \cdot 2 = 6t^2 = 6(2x-1)^2.$$

So, the differential of $(2x-1)^3$ is $6(2x-1)^2$.

Find the differential of $y = e^{x^2}$

Let $t = x^2$, so $\frac{dt}{dx} = 2x$.

$$y = e^t, \text{ so } \frac{dy}{dt} = e^t$$

$$\frac{dy}{dx} = \frac{dy}{dt} \cdot \frac{dt}{dx} = e^t \cdot 2x$$

But $t = x^2$, so $\frac{dy}{dx} = 2xe^{x^2}$.

20.2 The product rule

The **product rule** equation is: $\frac{dy}{dx} = u \cdot \frac{dv}{dx} + v \cdot \frac{du}{dx}$.

This is used to find the derivative of two or more functions multiplied together.

An example of two functions multiplied together is $y = 4x^2 (3x-1)^5$. One function is $4x^2$, the other is $(3x-1)^5$.

How it works

Let us consider $y = uv$, where u and v are functions of x. Making a small change in x, δx, gives rise to small changes, δy, δu and δv, in y, u and v respectively.

So:
$$y + \delta y = (u + \delta u)(v + \delta v) = uv + u\delta v + v\delta u + \delta u \delta v$$

Rearranging this gives:
$$\delta y = uv + u\delta v + v\delta u + \delta u \delta v - y$$

But $y = uv$, so:
$$\delta y = \cancel{uv} + u\delta v + v\delta u + \delta u \delta v - \cancel{uv}$$
$$\delta y = u\delta v + v\delta u + \delta u \delta v$$

So: $\dfrac{\delta y}{\delta x} = u\dfrac{\delta v}{\delta x} + v\dfrac{\delta u}{\delta x} + \dfrac{\delta u \delta v}{\delta x}$

When δx tends to zero, $\dfrac{\delta u \delta v}{\delta x}$ will also tend to zero, and

$$\frac{\delta y}{\delta x} \to \frac{dy}{dx}$$

$$\frac{\delta u}{\delta x} \to \frac{du}{dx}$$

$$\frac{\delta v}{\delta x} \to \frac{dv}{dx}$$

As $y = uv$, $\dfrac{dy}{dx} = \dfrac{d(uv)}{dx}$, resulting in the product rule equation: $\dfrac{dy}{dx} = u \cdot \dfrac{dv}{dx} + v \cdot \dfrac{du}{dx}$.

Using the product rule

Define u, and then calculate $\dfrac{du}{dx}$

Then define v, and calculate $\dfrac{dv}{dx}$.

Finally use all four parts in the product rule equation to calculate $\dfrac{dy}{dx}$.

EXAMPLE

Given that $y = 4x^2 (3x - 1)^5$, find $\dfrac{dy}{dx}$.

Let $u = 4x^2$, so $\dfrac{du}{dx} = 8x$

Also, let $v = (3x - 1)^5$

First we need to use the chain rule to calculate $\dfrac{dv}{dx}$.

Let $t = 3x - 1$, then $\dfrac{dt}{dx} = 3$.

Let $v = t^5$, then $\dfrac{dv}{dt} = 5t^4$.

$$\frac{dv}{dx} = \frac{dv}{dt} \cdot \frac{dt}{dx} = 5t^4 \cdot 3 = 15t^4 = 15(3x - 1)^4.$$

Substituting into the product rule equation gives:

$$\frac{dy}{dx} = u\frac{dv}{dx} + v\frac{du}{dx} = 4x^2 \cdot 15(3x - 1)^4 + (3x - 1)^5 \cdot 8x = 60x^2(3x - 1)^4 +$$

$8x(3x - 1)^5$

Taking out the common factor $4x(3x - 1)^4$ gives:

$$\frac{dy}{dx} = 4x(3x - 1)^4(15x + 2(3x - 1)).$$

Which can be simplified to:

$$\frac{dy}{dx} = 4x(3x - 1)^4(21x - 2).$$

20.3 The quotient rule

The **quotient rule** equation is:

$$\frac{dy}{dx} = \frac{v\dfrac{du}{dx} - u\dfrac{dv}{dx}}{v^2}.$$

The quotient rule is used to find the derivative where one or more functions is divided by one or more other functions.

An example is the equation $y = \frac{x^3 - 3}{x^2 + 5}$. Here $x^3 - 3$ is one function, which is divided by another function: $x^2 + 5$.

How it works

$y = \frac{u}{v}$ can be written as $y = uv^{-1}$.

Substituting v^{-1} for v in the product rule equation gives:

$$\frac{dy}{dx} = u\frac{d(v^{-1})}{dx} + v^{-1}\frac{du}{dx}$$

To calculate $\frac{d(v^{-1})}{dx}$, let $t = v^{-1}$, then $\frac{dt}{dv} = -v^{-2}$

So from the chain rule, $\frac{dt}{dx} = \frac{dt}{dv} \cdot \frac{dv}{dx}$,

$$\frac{d(v^{-1})}{dx} = -v^{-2}\frac{dv}{dx}.$$

Hence the product rule equation becomes:

$$\frac{dy}{dx} = -uv^{-2}\frac{dv}{dx} + v^{-1}\frac{du}{dx} = \frac{1}{v}\frac{du}{dx} - \frac{u}{v^2}\frac{dv}{dx}.$$

This gives the quotient rule equation:

$$\frac{dy}{dx} = \frac{v\frac{du}{dx} - u\frac{dv}{dx}}{v^2}.$$

Using the quotient rule

As $y = \frac{u}{v}$, u is defined already as the numerator and v as the denominator.

From u, calculate $\frac{du}{dx}$.

From v, calculate v^2, and $\frac{dv}{dx}$.

Then put these values in the quotient rule equation to calculate $\frac{dy}{dx}$.

EXAMPLE

Given that $y = \frac{x^3 - 3}{x^2 + 5}$ find $\frac{dy}{dx}$.

Let $u = x^3 - 3$, then $\frac{du}{dx} = 3x^2$

Let $v = x^2 + 5$, then $v^2 = (x^2 + 5)^2$, and $\frac{dv}{dx} = 2x$

So $\frac{dy}{dx} = \frac{v\frac{du}{dx} - u\frac{dv}{dx}}{v^2} = \frac{(x^2 + 5)\,3x^2 - (x^3 - 3)2x}{(x^2 + 5)^2} = \frac{x^4 + 15x^2 - 6x}{(x^2 + 5)^2}$

Therefore $\frac{dy}{dx} = \frac{-x^3 + x^2 - 6x}{x^2 + 5}$.

20.4 The differential of $\ln|x|$

$\dfrac{dx}{dy} \cdot \dfrac{dy}{dx}$ equals 1, by definition.

If $y = \ln x$, then $e^y = x$

Therefore $\dfrac{dx}{dy} \cdot \dfrac{dy}{dx} = \dfrac{d(e^y)}{dy} \cdot \dfrac{dy}{dx}$

Section 19.14 shows that the differential e^x with respect to x is e^x.

So $\dfrac{d(e^y)}{dy} = e^y$.

So $\dfrac{d(e^y)}{dy} \cdot \dfrac{dy}{dx} = e^y \cdot \dfrac{dy}{dx} = 1$.

Rearranging this gives $\dfrac{dy}{dx} = \dfrac{1}{e^y} = \dfrac{1}{x}$.

Therefore the differential of $\ln|x|$ with respect to x is $\dfrac{1}{x}$.

Test yourself

The answers are given on pages 214–15.

Question 20.1
Differentiate the following with respect to x:
a) $(3x^2 + 2x + 5)^2$
b) $y = x^3 (x - 2)^2$
c) $y = x^3 e^x$
d) $y = e^x \ln x$
e) $y = \dfrac{x^2 - 1}{x^2 + 1}$

Question 20.2
Given that $y = \dfrac{x}{x^2 + 1}$

a) Find $\dfrac{dy}{dx}$.

b) Find the values of x for which $\dfrac{dy}{dx} = 0$.

Question 20.3
The Mosteller equation for Body Surface Area (BSA) in humans is

$$BSA = \left(\dfrac{\text{height (cm)} \times \text{weight (kg)}}{3600}\right)^{\frac{1}{2}}$$

The equation for Body Mass Index (BMI) is

$$BMI = \dfrac{\text{weight (kg)}}{[\text{height (m)}]^2}$$

a) Combine these two equations to write BSA in terms of height and BMI.
b) A year ago, an adolescent was 150 cm tall and weighed 54 kg. During the year he has grown 10 cm and his BMI has remained constant. Assuming linear growth, what was the change in his BSA with respect to the change in his height halfway through the year? (Hint: differentiate your answer for part (a) of this question to make an equation for change in BSA with respect to height. Then plug in the appropriate values for the midpoint of the year.)

Integral calculus

Integral calculus is concerned with **integration**. Integration is the reverse process to differentiation.

For our purposes, there are two main uses of integration:

- to calculate the original equation for a curve when we only know its differential;
- to calculate the area underneath a portion of a curve.

21.1 Integration: the converse of differentiation

We learnt in Chapter 19 that the differential of $y = x^2$ is $\dfrac{dy}{dx} = 2x$.

So, the integral of $2x$ would appear to be x^2.

Simple. Except, we also learnt that the differential of $y = x^2 + 9$ is $dy/dx = 2x$, so the integral of $2x$ could also be $x^2 + 9$.

Thus, in the same way that we *lose* the constant (because the differential of the constant is zero) when differentiating, we have to *replace* the constant when integrating.

This unknown constant is symbolised by the capital letter C.

Therefore the integral of $2x$ is $x^2 + C$.

> **EXAMPLE**
>
> The integral of $46x$ is the same as the integral of $23(2x)$.
>
> The integral of $23(2x) = 23x^2 + C$.

Because we cannot give C as a number we call this the **indefinite integral**. The integral becomes **definite** if a co-ordinate is given, and C can then be calculated.

21.2 The integration symbol

The symbol to indicate integration is $\int$. Think of it as an elongated S for "sum up".

The $\int$ symbol is always followed by *d(something)*, in these examples *dx*, meaning with respect to *x*.

So, using the example in the previous section, the integral of $2x$ with respect to x is written as $\int 2x\ dx$, so

$$\int 2x\ dx = x^2 + C.$$

The symbol $\int_{2}^{5} f(x)dx$ means the integral of the function of x with respect to x between the values of $x = 2$ to $x = 5$.

Using function notation (i.e. $y = f(x)$) we can state $\int f'(x)\ dx = f(x)$ and we can see that integration is the opposite of differentiation. The differential of $f(x)$ with respect to x is $f'(x)$ (see Section 19.7) and the integral of $f'(x)$ with respect to x is the function $f(x)$.

21.3 The integral of x^n

The integral of x^n with respect to x is

$$\frac{x^{n+1}}{n+1} + C.$$

Using the integration symbol, this is written as

$$\int x^n\ dx = \frac{x^{n+1}}{n+1} + C$$

when $n \neq -1$.

Note that $\neq$ is the symbol for "not equal to".

EXAMPLE

We wish to integrate $y = 4x^3$ with respect to x.

$$\int 4x^3\ dx = 4\frac{x^{3+1}}{3+1} + C = 4\frac{x^4}{4} + C = x^4 + C$$

The integral is therefore $x^4 + C$.

21.4 The integral of $\frac{1}{x}$

The general equation for integrating x^n given in Section 21.3 also works for negative values of n (e.g. x^{-5}), except for when $n = -1$.

When $n = -1$, then $x^n = x^{-1} = \frac{1}{x}$.

In Section 20.4 the differential of $\ln x$ was shown to be $\frac{1}{x}$.

So when $n = -1$, $\int \frac{1}{x}\ dx = \ln|x| + C$.

21.5 The integral of e^x

In Section 19.14 the differential of e^x was shown to be e^x.

Thus $\int e^x = e^x + C$.

21.6 Calculating areas by using integration

With a straight-line graph, it is easy to calculate the area under part of the line.

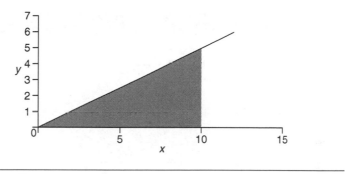

Graph of $y = x/2$

The equation for this graph is $y = \dfrac{x}{2}$.

The area under the line between the values $x = 0$ and $x = 10$ is calculated by:

$$\frac{\delta y \times \delta x}{2} = \frac{10 \times 5}{2} = 25$$

However, to calculate the area under a curve we need to use integration.

The area under a curve is the *integral* of the equation for the curve.

The area under the curve $y = x^n$ is the integral of x^n. We use the capital letter A to symbolise area.

A = integral with respect to x of x^n

$$A = \int x^n \, dx = \frac{x^{n+1}}{n+1} + C$$

This integral is defined by cancelling the C when two values of x are given as boundaries, in this case between $x = 0$ and $x = 10$.

For $y = \dfrac{x}{2}$

$$\int_0^{10} \frac{x}{2} \, dx = \left[\frac{x^{1+1}}{2(1+1)} + C \right]_0^{10} = \left[\frac{x^2}{4} + C \right]_0^{10}$$

Notice that square brackets are the standard notation when using the limits for definite integration.

As the numbers at the top and bottom of the square brackets define the limits, substitute $x = 10$ into the equation, and then again $x = 0$ into the equation. Then find the area between by subtracting one from the other.

$$\left(\frac{10^2}{4} + C\right) - \left(\frac{0^2}{4} + C\right) = 25 - 0 + C - C = 25$$

Note how the constant C is cancelled out. This is also known as "finding the definite integral", because there are no remaining unknown constants.

EXAMPLE

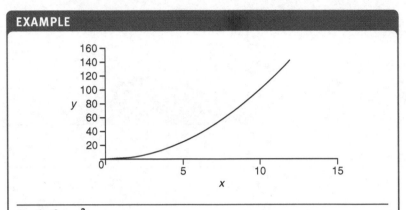

Graph of $y = x^2$

The equation for the area under the curve is the integral of x^2:

$$A = \int x^2 \, dx = \frac{x^{2+1}}{2+1} + C = \frac{x^3}{3} + C$$

If we want to calculate the area between $x = 0$ and $x = 10$ we can write this as

$$A = \int_0^{10} x^2 \, dx = \left[\frac{x^3}{3} + C\right]_0^{10} = \left(\frac{10^3}{3} + C\right) - \left(\frac{0^3}{3} + C\right) = \frac{10^3}{3} - 0$$

$= 333.33$ to two decimal places.

If we want to calculate the area between $x = 1$ and $x = 10$ we can write this as

$$A = \int_1^{10} x^2 \, dx = \left[\frac{x^3}{3} + C\right]_1^{10} = \left(\frac{10^3}{3} + C\right) - \left(\frac{1^3}{3} + C\right) = \frac{10^3}{3} - \frac{1}{3} = 333.\dot{3}.$$

21.7 How does it work?

This section explains the mathematical proof that the area under a curve is equal to the integral of the formula for the curve.

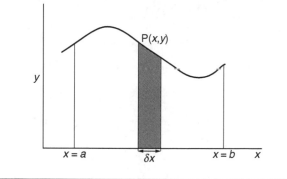

Graph showing $y \times \delta x$

The approximate area of the thin strip under the curve in this graph is the height of the strip, y, multiplied by the difference between the strip's two points on the x-axis, which can be written as δx.

So, the area of the strip approximates to

$y \times \delta x$

i.e. $y\,\delta x \approx \delta A$

Note the use of the $\approx$ sign, meaning "approximately equal to".

If we fill the area from $x = a$ to $x = b$ with an infinite number of infinitely small strips, and add their areas, the mathematical notation is

$$A \approx \sum_{x=a}^{x=b} \delta A$$

Remember A symbolises the area, and that the sigma sign $\sum$ means a summation of all that follows it. The $x = b$ and $x = a$ symbols above and below the $\sum$ sign mean that all the strips from $x = b$ to $x = a$ are included in the expression.

We've already stated that $\delta A \approx y\,\delta x$.

Substituting $y\,\delta x$ for δA in $A \approx \sum_{x=a}^{x=b} \delta A$ gives:

$$A \approx \sum_{x=a}^{x=b} y\,\delta x$$

As the width of the strips, i.e. δx, gets smaller, the estimation of the area becomes more accurate.

We've met the lim symbol already in Section 19.6.

When the strips are infinitely small, i.e. $\delta x \rightarrow 0$, we have an accurate equation for the area under the curve, i.e.

$$A = \lim_{\delta x \to 0} \sum_{x=a}^{x=b} y\,\delta x$$

As $y = f(x)$ then A is also a function of x alone.

Therefore we can manipulate the equation $\delta A \approx y\delta x$ to

$$\frac{\delta A}{\delta x} \approx y$$

This also becomes more accurate as δx gets smaller:

$$\lim_{\delta x \to 0} \frac{\delta A}{\delta x} = y$$

But $\lim_{\delta x \to 0} \frac{\delta A}{\delta x}$ is $\frac{dA}{dx}$ so $\frac{dA}{dx} = y$

If $\frac{dA}{dx} = y$, then

$$A = \int y\,dx$$

so the area is the integral of y with respect to x.

So, with limits $x = a$ and $x = b$ we can say:

$$\text{Total area } (A) = \int_{a}^{b} y\,dx$$

Thus integration can be seen as calculating the area under a curve between the limits of $x = a$ and $x = b$.

Test yourself

The answers are given on page 215.

Question 21.1
Integrate $y = 7x^4$ with respect to x.

Question 21.2
Find the area under the curve from $x = 3$ to $x = 5$ for the curve $y = 2x^3$ shown in the following graph.

Question 21.3
The rate, in calories per minute, at which a person using an exercise machine metabolises calories is modelled by the function:
$f(t) = -\frac{1}{4}t^3 + \frac{3}{2}t^2 + 1$, where t = exercise time in minutes.
How many calories are metabolised in the first four minutes on the exercise machine?

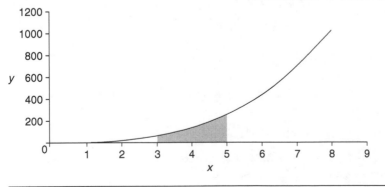

Graph of $y = 2x^3$

22 Further integral calculus

This chapter explains how to integrate more complicated functions.

In the same way that we use the chain, product and quotient rules to differentiate more complicated functions, we can use the opposite of these rules to *integrate* more complicated functions.

22.1 Integration by substitution

The principle of **integration by substitution** is that:

$\int u^n \frac{du}{dx} dx = \int u^n\, du$, when $n \neq -1$.

This is the chain rule worked backwards.

We use this if the function that we are trying to integrate is equal to a substitute function, u, multiplied by the substitute's derivative, $\frac{du}{dx}$.

An example of this is in the expression $\int \frac{x}{(x+1)^{1/2}} dx$, where we can use the substitute $u^2 = x + 1$ to simplify the calculation.

How it works

We use a substitute to change the equation algebraically, so that the integration becomes easier.

As our equation is now in terms of u, we now replace dx with $\frac{dx}{du} du$:

$$\int u^n \frac{du}{dx} dx = \int u^n \frac{du}{dx} \cdot \frac{dx}{du} du = \int u^n \cdot 1\, du = \int u^n\, du$$

Using integration by substitution

Define the substitute u in terms of x.

Rearrange this to find x in terms of u.

Calculate $\frac{du}{dx}$.

Use this to define the rest of the integral in terms of u.

Now calculate $\frac{dx}{du}$.

Then use this in the formula $\int u^n \frac{du}{dx} \cdot \frac{dx}{du} du$ to produce an integral in terms of u.

Next calculate the integral with respect to u.

Finally, rewrite the answer in terms of x rather than u.

EXAMPLE

Use the substitute $u^2 = x + 1$ to calculate $\int \dfrac{x}{(x+1)^{1/2}}\,dx$

As $u^2 = x + 1$, then $x = u^2 - 1$, so $\dfrac{dx}{du} = 2u$, and $(x+1)^{1/2} = u$

We can now rewrite our first integral in terms of u:

$$\int \frac{x}{(x+1)^{1/2}}\,dx = \int \frac{u^2 - 1}{u}\,dx$$

However, it is not possible to integrate a function of u with respect to x, so we replace dx with $\dfrac{dx}{du}\,du$.

However, because $x = u^2 - 1$ and $\dfrac{dx}{du} = 2u$, then

$$\int \frac{x}{(x+1)^{1/2}}\,dx = \int \left(\frac{u^2 - 1}{u}\right)\left(\frac{dx}{du}\right)du = \int \frac{u^2 - 1}{u}(2u)\,du = \int \frac{2u^3 - 2u}{u}\,du$$

$$= \int 2u^2 - 2\,du$$

Integrating this gives $\int \dfrac{x}{(x+1)^{1/2}}\,dx = \dfrac{2}{3}u^3 - 2u + C$

Now we revert to variable x by substituting in $u = (x+1)^{1/2}$:

$$\int \frac{x}{(x+1)^{1/2}}\,dx = \frac{2}{3}(x+1)^{3/2} - 2(x+1)^{1/2} + C.$$

To find the definite integral, i.e. the interval between two points, rewriting back to the original variable is not necessary.

EXAMPLE

Using the same example as above, use the substitute $u^2 = x + 1$ to calculate

$$\int_{0}^{8} \frac{x}{(x+1)^{1/2}}\,dx$$

Using $u^2 = x + 1$, when $x = 0$, $u = 1$, and when $x = 8$, $u = 3$

$$\int_{0}^{8} \frac{x}{(x+1)^{1/2}}\,dx = \int_{1}^{3} (2u^2 - 2)\,du = \left[\frac{2}{3}u^3 - 2u + c\right]_{1}^{3} =$$

$$(18 - 6 + c) - \left(\frac{2}{3} - 2 + c\right) = 12 - \left(-\frac{4}{3}\right) = 13\frac{1}{3}.$$

Note how the constant cancels itself out.

22.2   Integration by parts

The **integration by parts** equation is:

$$\int v \frac{du}{dx} dx = uv - \int u \frac{dv}{dx} dx.$$

This is the product rule worked backwards.

We use this to integrate a product of two functions, for example, $\int xe^x dx$.

How it works

The product rule for differentiation in Chapter 20 is as follows:

$$\frac{d(uv)}{dx} = u \cdot \frac{dv}{dx} + v \cdot \frac{du}{dx}$$

This can be rewritten as:

$$v \frac{du}{dx} = \frac{d(uv)}{dx} - u \frac{dv}{dx}$$

Integrating this gives the integration by parts equation:

$$\int v \frac{du}{dx} dx = uv - \int u \frac{dv}{dx} dx$$

Using integration by parts

Define v and $\frac{du}{dx}$, then calculate $\frac{dv}{dx}$ and u.

Then substitute into the integration by parts formula.

EXAMPLE

Find $\int x\, e^x\, dx$

Let $v = x$, so $\dfrac{dv}{dx} = 1$

$\dfrac{du}{dx} = e^x$, so $u = e^x$

Substituting into the integration by parts equation

$$\int v \frac{du}{dx} dx = uv - \int u \frac{dv}{dx} dx$$

gives $\int x\, e^x\, dx = x\, e^x - \int e^x \cdot 1\, dx = x\, e^x - e^x + C.$

There are occasions when integrating by parts needs to be repeated.

EXAMPLE

Find $\int x^2\, e^{2x}\, dx$

Let $v = x^2$, and $\dfrac{dv}{dx} = 2x$

Let $\dfrac{du}{dx} = e^{2x}$, therefore $u = \dfrac{1}{2}\, e^{2x}$

Substituting into the integration by parts equation gives

$$\int x^2 e^{2x}\, dx = \frac{1}{2}\, e^{2x} \cdot x^2 - \int \frac{1}{2}\, e^{2x} \cdot 2x\, dx = \frac{1}{2}\, x^2 e^{2x} - \int x e^{2x}\, dx$$

So we now need to find $\int x e^{2x}\, dx$.

Let $v = x$, therefore $\dfrac{dv}{dx} = 1$

As above, $\dfrac{du}{dx} = e^{2x}$, therefore $u = \dfrac{1}{2}\, e^{2x}$

Substituting a second time into the integration by parts equation gives

$$\int x e^{2x}\, dx = \frac{1}{2}\, e^{2x} \cdot x - \int \frac{1}{2}\, x e^{2x} \cdot 1 = \frac{1}{2}\, x e^{2x} - \frac{1}{4}\, e^{2x} + C$$

Now we return to the first equation and substitute into it:

$$\int x^2 e^{2x}\, dx = \frac{1}{2}\, x^2 e^{2x} - \frac{1}{2}\, x e^{2x} + \frac{1}{4}\, e^{2x} + C = \frac{1}{4}\, e^{2x}\,(2x^2 - 2x + 1) + C$$

If your calculation is getting more complicated rather than simpler, start again, changing what you use as v and which you use as $\dfrac{du}{dx}$.

22.3 The integral of e^{kx}

In Section 21.5 we found that $\int e^x = e^x + C$.

However, if a constant is involved, by using integration by parts this becomes the more general formula:

$$\int e^{kx}\, dx = \frac{1}{k}\, e^{kx} + C \qquad \cdot$$

22.4 The integral of $\ln|x|$

Let $v = \ln|x|$, and $\dfrac{dv}{dx} = \dfrac{1}{x}$

Let $\dfrac{du}{dx} = 1$, and $u = x$

Substituting these into the integration by parts formula gives the formula for the integration of $\ln|x|$ by parts:

$$\int \ln|x| \cdot dx = x \ln|x| - x + c$$

Test yourself

The answers are given on pages 215–216.

Question 22.1
Integrate the following using the substitution given:

a) $\int x(x^2 + 1)^3 \, dx$, using $u = x^2 + 1$

b) $\int xe^{x^2} \, dx$, using $u = x^2$

c) $\int x^3 \sqrt{(x^4 - 1)} \, dx$, using $u = x^4$

d) $\int_1^2 x(1 + 2x^2) \, dx$, using $u = 1 + 2x^2$

Question 22.2
Integrate $x^2 e^x$ using parts.

Question 22.3
Find the definite integral $\int_1^2 x^5 \ln x \, dx$ using parts.

Question 22.4
The area under a plasma concentration–time curve (known as AUC, Area Under Curve), reflects the body's exposure to a drug after the oral administration of a dose of that drug. It is expressed in mg h l^{-1}.
For a drug being studied, in the first 2 hours after administration of the drug its plasma concentration can be described by the equation $y = 4x$, where x is the time since administration in hours and y is the plasma concentration in mg l^{-1}.
From 2 to 12 hours after the drug has been administered, it can be described by

$y = 8e \left(\dfrac{2}{5} - \dfrac{x}{5} \right)$, where e is the exponential constant ≈ 2.718.

This can be drawn in a graph:

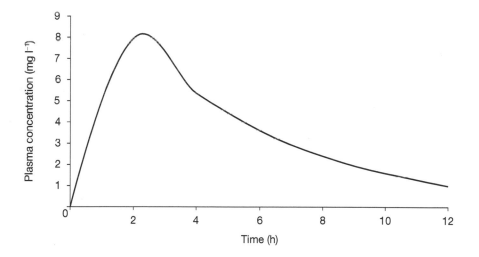

Calculate the body's exposure to the drug in mg h l^{-1}, i.e. the area under the curve. (Hint: use integration to calculate the area under the first part of the curve, and integration by substitution for the second part.)

Using graphs

Different types of data need different types of graph. This chapter explains which graphs are suitable and how to interpret them.

23.1 Labelling graphs

Some general comments on plotting graphs:

- always label the axes with the names of the variables;
- also, where relevant, give the units used in brackets after the label, for example: Time (min); and
- give the graph a title.

23.2 Scatter plots

Chapter 8 explained the rudiments of plotting a graph from a table of data.

One way to decide whether there is a relationship between two sets of data is by drawing a **scatter plot,** a graph showing all the data as individual points.

We can use a scatter plot when the units on the x- and y-axes are **continuous,** i.e. when we can use values in between the co-ordinates that we are plotting.

Examples of continuous variables are time and distance.

If there seems to be a pattern, we can join the points on the plot.

The following table and its graph are an example of a **frequency distribution,** the arrangement of the values of one or more variables in a sample. Each entry in the table contains the number (the "frequency") of occurrences which are "distributed" within each particular timeframe.

EXAMPLE

This table gives the number of bacteria growing in a colony as a function of time.

Table showing growth of bacterial colony								
Time (min)	0	10	25	45	60	75	100	120
No. of bacteria	470	650	1030	1900	3040	4830	10 500	19 500

The data are effectively continuous:

- time is certainly continuous;

- although we can't have half a bacterium, the numbers of bacteria are so large that that they can be treated as being continuous.

Thus we can use a scatter plot and, as there is clearly a pattern, draw a line through the points.

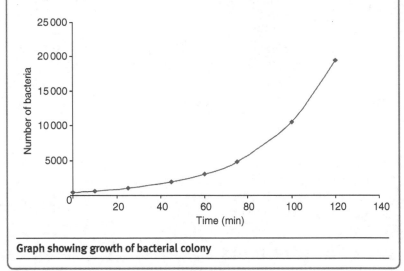

Graph showing growth of bacterial colony

23.3 Graphs and types of variable

Because continuous variables are those which can take any value within a given range, we can get information from the line that we have drawn. In the example above, we can estimate the number of bacteria in the colony after 90 minutes even though no data were taken at that point.

For **categorical variables**, those where only certain values can exist, or **nominal variables**, those that have no ordering to their categories, we cannot join the points on a scatter plot.

Similarly, if the data are **discrete**, meaning that the values in between the co-ordinates are meaningless, we cannot join the points on a scatter plot.

EXAMPLE

The number of patients admitted to a hospital because of fractures is collected over 7 consecutive days.

Table of patients admitted to hospital over 1 week							
Day	1	2	3	4	5	6	7
Number of admissions due to fractures	5	4	6	7	8	5	4

As it is not possible to have a half a patient, the data are discrete, so we can use a scatter plot but we cannot join the points.

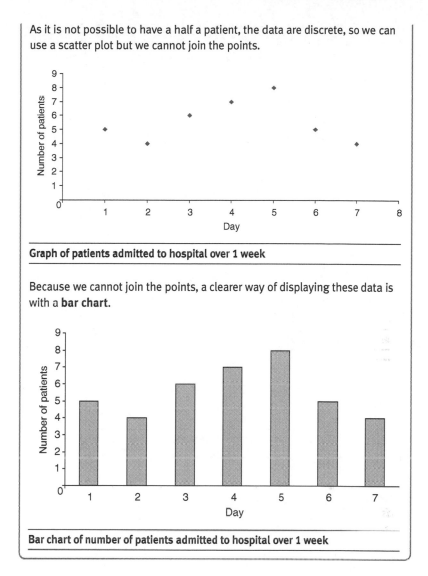

Graph of patients admitted to hospital over 1 week

Because we cannot join the points, a clearer way of displaying these data is with a **bar chart**.

Bar chart of number of patients admitted to hospital over 1 week

23.4 Histograms

A **histogram** is another way of showing the distribution of a continuous variable. The data are shown as adjacent rectangles, each with an area equal to the frequency of the observations in that interval.

In the following example, some of the data related to a single year, some to two years, and some to three. So, in the graph of the data, the height of each rectangle equals the frequency divided by the width of the interval (one, two or three years).

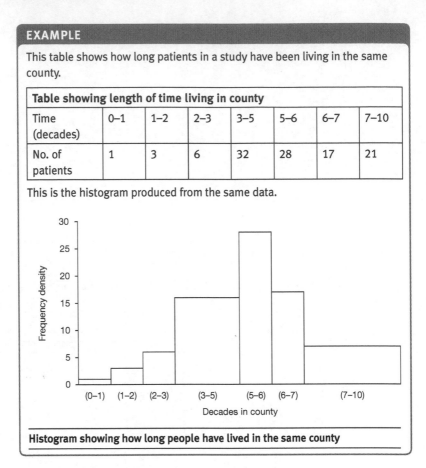

EXAMPLE

This table shows how long patients in a study have been living in the same county.

Table showing length of time living in county							
Time (decades)	0–1	1–2	2–3	3–5	5–6	6–7	7–10
No. of patients	1	3	6	32	28	17	21

This is the histogram produced from the same data.

Histogram showing how long people have lived in the same county

23.5 Survival curves

Data for survival of organisms or patients over time are cumulative, so it is often more meaningful to use a **cumulative frequency** chart.

EXAMPLE

Cumulative frequency of deaths of 100 patients from time of treatment for lung cancer							
Time (months)	<3	4–6	7–9	10–12	13–15	16–18	19–21	22–24
No. of deaths (cumulative)	26	46	61	71	78	81	82	82

Though researchers count the number of deaths in each time period, the resulting cumulative distribution is more often shown as a **survival curve**.

As time to death is a continuous variable, we can draw a line to connect the frequencies. The resulting survival curve can be used to make estimations, for instance of the median survival time: the time at which 50% of the population being studied is still alive. It may also give us an indication of whether the survival curve is "levelling off", giving us the proportion of patients who are surviving long-term.

The graph below shows the same data as a survival curve.

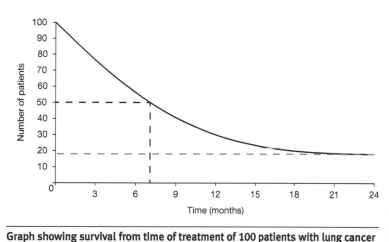

Graph showing survival from time of treatment of 100 patients with lung cancer

It can be estimated from the survival curve that 50% of the patients were still alive at 7 months, and that 18% were surviving long-term.

23.6 The line of best fit

When we collect and plot biological data, they don't usually follow an *exact* mathematical relationship (for instance, a straight line). This could be because of normal variations in the population or errors in collecting the data.

However, if the data seem close to a mathematical relationship, then we can demonstrate that relationship by drawing a **line of best fit** on the graph, rather than joining all the points.

The line of best fit is the line that best shows the trend of the plotted points.

If this is a straight line, we can use it to calculate the gradient.

EXAMPLE

The tail and body length of a group of 10 mice are related as follows:

Table showing relationship between body and tail length in 10 mice										
Body length (mm)	92	97	96	99	100	111	109	115	120	122
Tail length (mm)	31	32	35	36	40	43	44	49	49	52

Drawing a scatter plot of these data shows the relationship between the body and tail length.

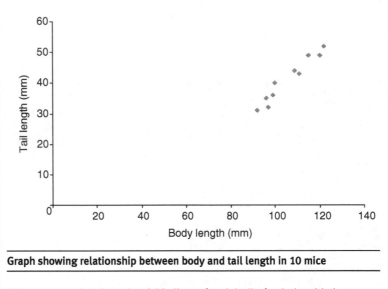

Graph showing relationship between body and tail length in 10 mice

This suggests that there is a fairly linear (straight-line) relationship between the length of the body of a mouse and its tail, and therefore it follows the $y = mx + c$ pattern.

EXAMPLE

Drawing the line of best fit to the scatter plot above gives the following.

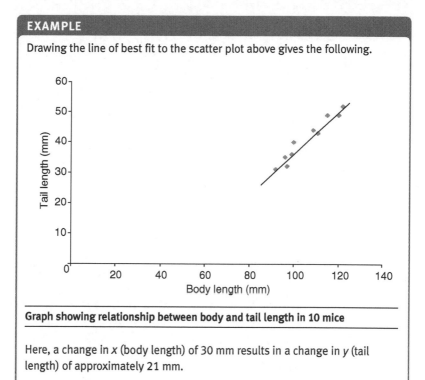

Graph showing relationship between body and tail length in 10 mice

Here, a change in x (body length) of 30 mm results in a change in y (tail length) of approximately 21 mm.

Remember that

$$\text{Gradient} = \frac{\text{Change in } y}{\text{Change in } x}$$

or

$$\text{Gradient} = \frac{y_2 - y_1}{x_2 - x_1}$$

So here, the gradient is approximately

$$\frac{21}{30} = 0.7$$

so the equation for the graph is, $y \approx 0.7x + c$

If we continue the line of best fit until it crosses the *y*-axis we can also estimate the constant *c*.

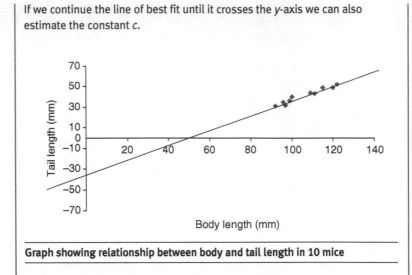

Graph showing relationship between body and tail length in 10 mice

The line crosses the *y*-axis at $y \approx -35$, so we can now write out the equation that links body and tail length in this group of mice:

$$y \approx 0.7x - 35$$

Obviously, we cannot have mice with negative tail lengths, so we can only predict tail length within the range of the available data.

Chapter 45 explains an exact way of calculating the line of best fit.

Recognising patterns in graphs

Many equations have characteristic curves – drawing a graph and recognising the curve can help us decide what the relationship between two variables is.

24.1 The quadratic relationship

In Chapter 13, we learnt that **quadratic equations** have two solutions, and that they have the formula:

$$ax^2 + bx + c = 0$$

They have a symmetrical $\cup$ shape (or, for negative values like $-ax^2$, a $\cap$ shape).

EXAMPLE

This graph does have a symmetrical $\cup$ shape so there may well be a quadratic relationship between x and y ...

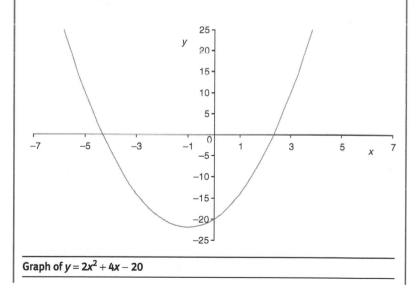

Graph of $y = 2x^2 + 4x - 20$

... and indeed there is. It is a graph of the quadratic equation:

$$y = 2x^2 + 4x - 20$$

Note that, as expected, the line crosses the y-axis at -20, the constant of the equation.

24.2 A cubic relationship

A polynomial with a **cube** as the highest power results in a characteristic tilted 2 or S shape.

EXAMPLE

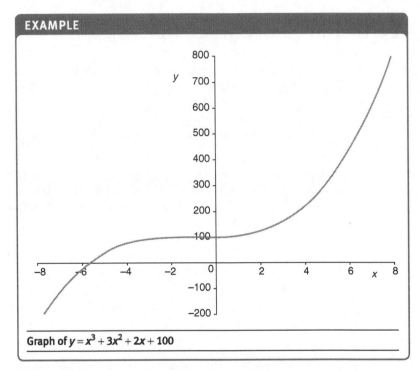

Graph of $y = x^3 + 3x^2 + 2x + 100$

24.3 Graphs suggesting asymptotes

An **asymptote** is where a function gets infinitely close to a line without crossing it.

Plotting a graph that appears to have asymptotes gives information about the relationship between the two variables.

We will meet the **Michaelis–Menten** equation for enzyme-catalysed reactions in Section 29.3.

At this stage, just note the pattern of its graph of substrate concentration, [S], against rate of reaction, V:

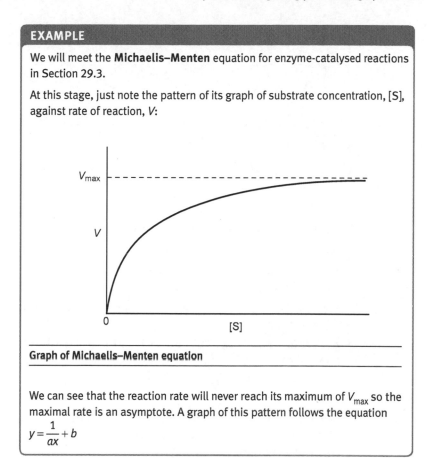

Graph of Michaelis–Menten equation

We can see that the reaction rate will never reach its maximum of V_{max} so the maximal rate is an asymptote. A graph of this pattern follows the equation

$$y = \frac{1}{ax} + b$$

24.4 Exponential relationships

The chapter on **exponential** relationships showed that exponential growth can be described by the equation $y = a^x$.

The shape of an exponential growth graph is characteristic: the line continues to go up and to the right forever.

Twenty seeds of rye grass are sown and grown under optimal conditions, and the numbers of grass seedlings is counted every month.

Table of number of grass seedlings over time							
Time (months)	0	1	2	3	4	5	6
No. of grass seedlings	20	35	61	105	188	320	557

Plotted on a graph, we get the following curve:

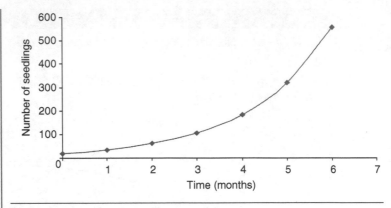

Graph of number of grass seedlings over time

The graph looks as if it could be exponential. To assess this, the graph can be plotted either:

- on "semi-log" graph paper, where one scale is logarithmic, and the other scale is not; or
- as a graph of time against the natural logarithm of the number of seedlings, symbolised by ln(number of seedlings).

If, as we suspect, the relationship is exponential, the plot should be **transformed** (converted) into a straight line.

In the following graph, note the scale of the *y*-axis: each unit is 10 times that of the unit below, so it is a logarithmic scale. The *x*-axis is a normal, non-logarithmic scale. The graph paper is therefore semi-logarithmic.

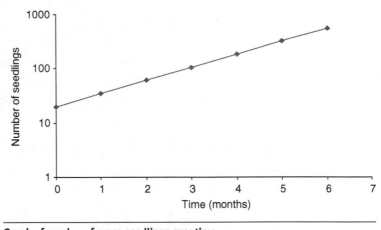

Graph of number of grass seedlings over time

Sure enough, the semi-logarithmic plot shows a straight line, confirming the exponential relationship.

We can use the same techniques to look for exponential decay.

> **EXAMPLE**
>
> Chicken meal is heated to 60°C to remove *Staphylococcus aureus* bacteria. The following are measurements of the surviving bacteria as a function of time:
>
> | Table showing bacterial numbers in chicken meal over time | | | | |
> |---|---|---|---|---|
> | Time (min) | 0 | 3 | 6 | 9 | 12 |
> | Bacteria (cells mm^{-3}) | 3×10^6 | 8.4×10^5 | 1.9×10^5 | 5.4×10^4 | 1.4×10^4 |

This produces the following curve:

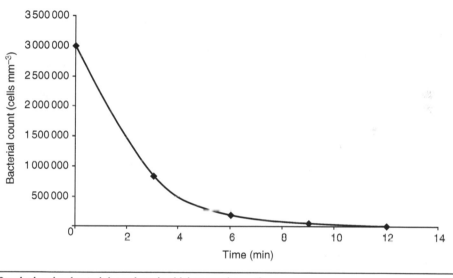

Graph showing bacterial numbers in chicken meal over time

We want to know whether this is a negative exponential.

This time, instead of using semi-log graph paper, we will convert the bacterial counts to their natural logarithms, i.e. ln (bacterial count).

Table of natural log of bacterial count in chicken meal over time					
Time (minutes)	0	3	6	9	12
Bacteria (cells mm^{-3})	3×10^6	8.4×10^5	1.9×10^5	5.4×10^4	1.4×10^4
ln(bacterial count)	14.91	13.64	12.15	10.90	9.55

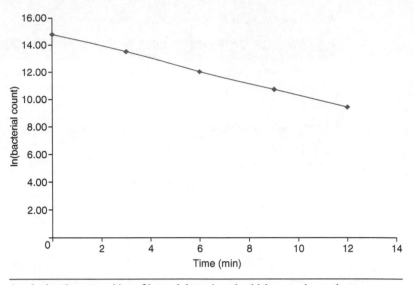

Graph showing natural log of bacterial numbers in chicken meal over time

The plot is now straight and has a negative gradient, confirming that the relationship between the bacterial count and time is one of exponential decay.

SI units

Any data that we use should, ideally, be converted to **SI units** (Système International d'Unités) before working with them.

The base SI units most frequently used by life and medical scientists are shown with their symbols in the following table.

Table of the base SI units		
Quantity	SI unit	Abbreviation
Mass	kilogram	kg
Length	metre	m
Time	second	s
Temperature	kelvin	K
Amount of substance	mole	mol

All other SI units are derived from these.

25.1 SI units for large and small numbers

We can use a **metric prefix** to describe very large or small numbers, as an alternative to expressing them in powers of 10. Each prefix indicates a multiplication factor of 1000.

However, for calculations, numbers need to be converted to SI units, using **standard form** (explained in Section 5.2).

Table of metric prefixes and abbreviations		
Metric prefix	Abbreviation	Power of 10
exa-	E	10^{18}
peta-	P	10^{15}
tera-	T	10^{12}
giga-	G	10^{9}
mega-	M	10^{6}
kilo-	k	10^{3}

Metric prefix	Abbreviation	Power of 10
milli-	m	10^{-3}
micro-	μ	10^{-6}
nano-	n	10^{-9}
pico-	p	10^{-12}
femto-	f	10^{-15}
atto-	a	10^{-18}

EXAMPLES

$12\,\text{km} = 1.2 \times 10^{4}\,\text{m}$

$0.78\,\text{nm} = 0.78 \times 10^{-9}\,\text{m} = 7.8 \times 10^{-10}\,\text{m}$

For mass, SI prefixes are used with the gram, because the kilogram already has a prefix as part of its name:

$1\,\text{g} = 1 \times 10^{-3}\,\text{kg}$

$1.37\,\mu\text{g} = 1.37 \times 10^{-6}\,\text{g} = 1.37 \times 10^{-9}\,\text{kg}$

25.2 Use of non-SI units

Some units are accepted for use with the SI, including the following non-SI units.

Table of commonly used non-SI units		
Name	Symbol	Value in SI units
degree Celsius	°C	$0°C = 273.15\,\text{K}$
litre	l	$1\,\text{l} = 10^{-3}\,\text{m}^3$
angstrom	Å	$1\,\text{Å} = 10^{-10}\,\text{m}$
minute (time)	min	$1\,\text{min} = 60\,\text{s}$
hour	h	$1\,\text{h} = 3600\,\text{s}$
day	d	$1\,\text{d} = 86\,400\,\text{s}$

The formula to convert from °C to K is °C = K −273.15. However, note that a change in temperature of 1°C is the same as a change in temperature of 1 K.

Test yourself

The answers are given on page 216.

Question 25.1
An experiment to calculate the mass of DNA in a human cell gives a value of 5.5 pg. Give this value in SI units in standard form.

Question 25.2
An epidemiologist studies the population in an area 5 km long by 6 km wide. Give the area in SI units, in standard form.

Question 25.3
Some frozen cells are stored at −40.5°C. Give this temperature in kelvin.

Question 25.4
An experiment shows that a human eye is sensitive to wavelengths from 4000 Å. Give this measurement in SI units in (a) standard form and (b) using a metric prefix.

 # Moles

The mole is a key unit of biochemical quantity.

26.1 Molecular mass

The **molecular mass**, symbol MM, of a substance is the mass of one molecule expressed in **daltons**, symbol Da. A dalton is one-twelfth the weight of an atom of ^{12}C, also known as carbon-12, the most common naturally occurring isotope of the element carbon.

Atomic mass is the mass of one atom expressed in daltons. It therefore also follows that a molecular mass is the sum of the relevant atomic masses.

EXAMPLE

Atomic mass of sodium (Na)	22.99 Da
Atomic mass of chlorine (Cl)	+ 35.45 Da
Molecular mass of sodium chloride (NaCl)	58.44 Da

26.2 Relative molecular mass

The **relative molecular mass** (symbol M_r), is the ratio of the molecular mass to the mass of one-twelfth the weight of an atom of ^{12}C. Thus, M_r is a pure number with no units.

EXAMPLE

The molecular mass of the enzyme trypsin is 23 300 Da, abbreviated to 23.3 kDa. Its relative molecular mass M_r, is therefore 23 300.

26.3 Moles

The strict definition of a **mole** is the amount of a substance that contains the same number of atoms, molecules, ions, or other elementary units as the number of atoms in 0.012 kg of ^{12}C. Its symbol is "mol".

0.012 kg of carbon-12 contains 6.022×10^{23} particles, so a mole of any substance also contains 6.022×10^{23} particles.

The value 6.022×10^{23} is known as **Avogadro's number.**

Most frequently in the life and medical sciences, we refer to a mole in the context of molecules, so a mole of, say, glucose contains 6.022×10^{23} molecules. One mole of a compound has a mass equivalent to its relative molecular mass in grams.

> **EXAMPLE**
>
> The relative molecular mass of glucose is 180.18 so a mole of glucose weighs 180.18 g.

26.4 Calculating the number of moles of a sample

The formula to calculate the number of moles of a sample is:

$$\text{number of moles} = \frac{\text{mass of sample (g)}}{\text{relative molecular mass}}$$

> **EXAMPLE**
>
> 360.36g of glucose is
>
> $$\frac{360.36\,\text{g}}{180.18} = 2 \text{ mol.}$$

26.5 Molarity

The **concentration** of a solution is defined as the amount of the substance in a set volume:

$$\text{concentration} = \frac{\text{amount}}{\text{volume}}$$

The **molarity**, M, of a solution is its concentration expressed as the number of moles per litre.

$$M = \frac{\text{no. of moles}}{\text{volume (l)}} = \frac{\text{mass (g)}}{\text{relative molecular mass} \times \text{solution volume (l)}} \text{ mol}\,l^{-1}$$

When a solution contains 1 mole of a compound in 1 litre of solution, it is said to be a "1 molar" (1 M) solution.

> **EXAMPLE**
>
> A 1 M solution of NaCl contains $58.44\,\text{g}\,l^{-1}$.

26.6 Preparing solutions of known molarity

We can calculate the mass of a substance required to prepare a solution of given molarity as follows:

Required mass of substance (g) = relative molecular mass × molarity (M)
× required volume of solution (l)

> **EXAMPLE**
>
> To prepare 500 ml (i.e. 0.5 litres) of a 3 M solution of NaCl:
>
> mass of NaCl needed $= 58.44 \times 3\,M \times 0.5\,l = 87.66\,g$
>
> While the mole is an SI unit, for historical reasons moles and molarity are derived from grams and litres, which are not SI units.

26.7 Diluting molar solutions

Calculations for diluting stock solutions can be made using this equation:

$$\frac{\text{required molarity of new solution (M)}}{\text{molarity of stock solution (M)}} = \frac{\text{required volume of stock solution (l)}}{\text{required total volume of new solution (l)}}$$

> **EXAMPLE**
>
> From a 2 M copper sulphate ($CuSO_4$) solution, we wish to prepare 200 ml of 0.4 M $CuSO_4$ solution.
>
> $$\frac{0.4\,M}{2\,M} = \frac{\text{required volume (ml)}}{200\,ml}$$
>
> $$\text{Required volume} = \frac{0.4 \times 200}{2} = \frac{80}{2} = 40\,ml$$
>
> 40 ml of 2 M $CuSO_4$ solution, made up to 200 ml with water, will produce a 0.4 M solution.

Another approach to calculating stock dilutions is to use the following equation:

$$C_1 V_1 = C_2 V_2$$

where C_1 is the concentration of the stock solution, V_1 is the required volume of stock solution, C_2 is the required concentration of the new solution, and V_2 is the required volume of the new solution.

EXAMPLE

To calculate the amount of 10 M sodium hydroxide (NaOH) solution needed to make 50 ml of 2 M solution:

$$10\,M \times V_1 = 2\,M \times 50\,ml$$

$$V_1 = \frac{2 \times 50}{10} = 10\,ml$$

We therefore need 10 ml of 10 M NaOH solution, made up to 50 ml with water.

26.8 Per cent solutions

When the term **per cent** (symbol "**%**") is applied to a solution, this may be **% w/w** (per cent weight/weight), **% w/v** (per cent weight/volume) or **% v/v** (per cent volume/volume).

If this is not defined, it usually represents % w/v, which is the number of grams per 100 ml.

EXAMPLE

To calculate the concentration of a 2 M NaCl solution as a % w/v solution, we first need to know the relative molecular mass of NaCl, which is 58.44.

The amount of NaCl in 1 litre of a 2 M NaCl solution is therefore:

$$2 \times 58.44 = 116.88\,g$$

The amount in 100 ml is:

$$116.88 \times \frac{100}{1000} \approx 11.7\,g$$

A 2 M NaCl solution is thus equivalent to a 11.7% w/v NaCl solution.

We can also convert a per cent value to molarity.

EXAMPLE

To express a 5% w/v NaCl solution by its molarity, first we calculate the mass per litre.

$$5\%\ w/v\ NaCl = \frac{5\,g}{100\,ml} = 50\,g\,l^{-1}$$

The relative molecular mass of NaCl is 58.44.

$$50\,g\,l^{-1} = \frac{50\,g}{58.44} \approx 0.86\,M$$

A 5% NaCl solution is therefore equivalent to a 0.86 M solution.

26.9 Normality

A **normal** solution (1N, which can also be written as 1 mol l^{-1}) has one mole of hydrogen ions (H$^+$) for an acid, or for an alkali hydroxide ions (OH$^-$) in solution per litre.

Normality and molarity are related as follows:

$$N = n\,M$$

where N is the normality, n is the number of H$^+$ or OH$^-$ ions per molecule, and M is the molarity.

> **EXAMPLE**
>
> Sulphuric acid has two replaceable H$^+$ ions per H_2SO_4 molecule, so to calculate the normality of a 3 M sulphuric acid solution:
>
> $$\text{Normality} = n\,M = 2 \times 3 = 6\,N$$
>
> A 3 M sulphuric acid solution therefore has a normality of 6 N.

Test yourself

The answers are given on pages 216–17.

Question 26.1
Carbon dioxide has the formula CO_2. The atomic masses of these atoms are:
Carbon (C) 12.01 Da
Oxygen (O) 16.00 Da
What is the molecular mass of carbon dioxide?

Question 26.2
The relative molecular mass of ascorbic acid is 176.12. How many moles are there in 528.36 g of ascorbic acid?

Question 26.3
Uric acid is $C_5H_4N_4O_3$
The component atomic masses are as follows:
Carbon (C) 12.01 Da
Hydrogen (H) 1.01 Da
Nitrogen (N) 14.01 Da
Oxygen (O) 16.00 Da
What is the molecular mass of uric acid?

Question 26.4
The relative molecular mass of a substance is 300. How many molecules does 1.5 kg of the substance contain?

Question 26.5
The molecular mass of adenosine triphosphate (ATP) is 507.18 Da. It is estimated that the human body contains 250 g of ATP. To two significant figures, how many moles is that?

Question 26.6
The molecular mass of glucose is 180.18 Da. What mass of glucose is there in 200 ml of a 0.5 M solution?

Question 26.7
The molecular mass of urea is 60.06 Da. What mass of urea is there in 100 ml of an 8 M solution?

Question 26.8
How much 0.25 M solution can be made from 500 ml of stock 1 M NaCl solution?

Question 26.9
How much water needs to be added to 100 ml of stock 8 M urea solution to dilute it to 5 M?

Question 26.10
The molecular mass of glucose is 180.18 Da. What is the molarity of a 5% w/v solution?

Question 26.11
The molecular mass of urea is 60.06 Da. What is the molarity of a 40% w/v solution, to three significant figures?

Question 26.12
Sodium hydroxide (NaOH) has one replaceable OH$^-$ ion per molecule. Its relative molecular mass is 40. If 20 g of sodium hydroxide is made up to 1 l with water, what is the normality of the resulting solution?

 pH

A measure of hydrogen ion concentration, pH is a scale of acidity.

27.1 pH and the concentration of hydrogen ions

In water, at any one moment a number of water molecules are splitting ("dissociating") to produce hydroxide (OH^-) and hydrogen (H^+) ions. Almost immediately they recombine ("reassociate") back to H_2O.

This is represented by the following equation:

$$H_2O \rightleftharpoons H^+ + OH^-$$

In fact, the hydrogen ions actually join other H_2O molecules to form oxonium ions (H_3O^+):

$$H_2O + H_2O \rightleftharpoons H_3O^+ + OH^-$$

However, it is conventional to think of the H^+ ion as being a dissociation product of water, rather than the H_3O^+ ion.

pH is a measure of the concentration of H^+ ions in a solution. It is a logarithmic scale of hydrogen ion concentration and is defined by:

$$pH = -\log[H^+]$$

Note that the square brackets indicate concentration measured in molarity, so $[H^+]$ symbolises the molarity of hydrogen ions.

> **EXAMPLE**
>
> The concentration of H^+ ions in pure water is 10^{-7} M
>
> $$pH = -\log[H^+] = -\log 10^{-7} = -(-7) = 7$$
>
> Thus the pH of pure water is 7.

27.2 The ion product

In water, the product of the concentrations of hydrogen ions and hydroxide ions always remains the same:

$$[H^+][OH^-] = 10^{-14}$$

This constant is known as the **ion product of water**, K_w and is helpful when determining the pH of alkaline solutions.

> **EXAMPLE**
>
> Sodium hydroxide, NaOH, is a very strong base that is almost completely ionised to OH^- ions.
>
> In 1 M NaOH the concentration of OH^- is therefore also 1 M, so $[OH^-] = 1$.
>
> Substituting this into $[H^+][OH^-] = 10^{-14}$ gives:
>
> $$[H^+][OH^-] = [H^+] \times 1 = 10^{-14}$$
>
> So
>
> $$[H^+] = 10^{-14}$$
> $$pH = -log[H^+] = -log 10^{-14} = 14$$
>
> Thus the pH of 1 M NaOH is 14.

27.3 Acids and bases

There are various definitions of acids, but for our purposes the most useful one is that acids are substances that dissociate to produce hydrogen ions. Using this definition we can write the equation:

$$HA \rightleftharpoons H^+ + A^-$$

where HA is the acid and A^- is its conjugate base.

The **equilibrium constant** for this reaction, known as the **acid dissociation constant**, is K_a:

$$K_a = \frac{[H^+][A^-]}{[HA]}$$

Strong acids have a high K_a as they reach equilibrium when they are fully dissociated (fully "ionised"). Weak acids have a lower proportion of ionised molecules, so have a lower K_a.

Factors like the temperature of a solution affect the dissociation constants of acids and bases. Tables usually give the dissociation constant at 25°C.

K_a is a better measure of the "strength" of an acid than pH: adding more water to an acid solution will not change the value of the equilibrium constant K_a but it will change the H^+ ion concentration on which pH depends. The lower the dissociation constant, the "stronger" the acid.

The pK_a of an acid is defined as the negative logarithm of K_a:

$$pK_a = -log_{10} K_a.$$

EXAMPLE

The K_a of acetic acid is $1.78 \times 10^{-5}\,mol\,l^{-1}$

Its pK_a is:

$$-\log K_a = -\log(1.78 \times 10^{-5}) = 4.75$$

The equation for a base, B^-, which accepts H^+ ions, is:

$$B^- + H^+ \rightleftharpoons BH$$

where B^- is the base, and BH its conjugate acid.

Here, the equilibrium constant K_a is represented by:

$$Ka = \frac{[H^+][B]}{[BH^+]}$$

Again, $pK_a = -\log K_a$.

Test yourself

The answers are given on page 217.

Question 27.1
Hydrochloric acid, HCl, nearly fully dissociates to H^+ and Cl^- ions. What is the pH of 0.1 M HCl?

Question 27.2
The pH of a solution is 2. What is the concentration of H^+ ions?

Question 27.3
What is the pH of a 0.1 M solution of NaOH?

Question 27.4
The pH of an aqueous solution is 8.5. To three significant figures, and using a calculator, what is the concentration of OH^- ions?

Question 27.5
The pK_a of ammonia is 9.25. What is its K_a? You will need a calculator for this.

Question 27.6
What is the resulting pH when 1 ml of 1 M sodium hydroxide is added to 1 l of water? Give your answer to two significant figures.

 **Buffers**

A pH **buffer** is a solution that keeps the pH at a given value, resisting pH change even when acids or bases are added.

28.1 Making a buffer

There are two ways to make a buffer.

One way is by partially neutralising a weak acid (or base) with a strong base (or acid). This is called a **titration**.

The other way to make a buffer is to mix a weak acid (or base) with its conjugate acid (base).

28.2 Calculating the pH of a buffered solution

The **Henderson–Hasselbalch equation** relates pH to the composition of buffer solutions. The equation is:

$$pH = pK_a + \log\frac{[A^-]}{[HA]}$$

where $[A^-]$ is the concentration of the conjugate base, and $[HA]$ is the concentration of the acid.

When the number of conjugate acid molecules equals the number of conjugate base molecules, i.e. when $[A^-] = [HA]$, then $\frac{[A^-]}{[HA]} = 1$. As $\log 1 = 0$, the pH will be the same as the pK_a.

We can use the Henderson–Hasselbalch equation to calculate the pH of a buffer solution.

> **EXAMPLE**
>
> We wish to make a buffer by mixing Tris hydrochloride (Tris–HCl, the conjugate acid) with Tris base (the conjugate base).
>
> The pK_a of Tris–HCl is 8.3.
>
> If we mix 250 ml of 1 M Tris–HCl with 750 ml of 1 M Tris base, the resulting acid and base molarities are 0.25 M and 0.75 M.
>
> $$pH = pK_a + \log\frac{[A^-]}{[HA]} = 8.3 + \log\frac{0.75}{0.25} = 8.3 + 0.4\dot{7} = 8.\dot{7}.$$
>
> Thus the pH of the resulting buffer is 8.8, to two significant figures.

28.3 Calculating the titration needed for a given pH

We can also use the Henderson–Hasselbalch equation to calculate the titration needed to achieve a given pH.

EXAMPLE

Acetic acid is a weak acid with a pK_a of 4.75. Its conjugate base is the acetate ion. When the two are mixed, we get an "acetate buffer".

We wish to prepare an acetate buffer of pH 5.0

We use the Henderson–Hasselbalch equation to calculate the necessary ratio of the acid and its conjugate base.

$$\text{Desired pH} = 5.0 = 4.75 + \log\frac{[A^-]}{[HA]}$$

So,

$$\log\frac{[A^-]}{[HA]} = 5.0 - 4.75 = 0.25$$

$$\frac{[A^-]}{[HA]} = 1.778$$

We therefore need the ratio of acetate and acetic acid to be 1.8 to 1, to two significant figures.

Note that the Henderson–Hasselbalch equation does not tell us the exact amounts of acetic acid and acetate ion to use, only their ratio.

In this example, the ratio of 1.8 base to its acid could be made by mixing 1.8 litres of 1 M sodium acetate with 1 litre of 1 M acetic acid.

Test yourself

You will need to use your calculator to answer these questions.
The answers are given on pages 217–18.

Question 28.1
The pK_a of Tris–HCl is 8.3. If we mix 300 ml of 1 M Tris–HCl with 200 ml of 1 M Tris base, what is the resulting pH?

Question 28.2
Equal volumes of 0.2 M acetic acid and 0.1 M sodium acetate (the conjugate base) solutions are mixed to make a buffer. The pK_a of acetic acid is 4.75. Use a calculator to work out the pH of the resulting solution, to two significant figures.

Question 28.3
Diethylmalonic acid has a pK_a of 7.2. If we have 1 M diethylmalonic acid and 1 M conjugate base, how much of each do we need to make 1 litre of pH 7.4 buffer?

Question 28.4
An acetic acid buffer has a pH of 4.8. The K_a for acetic acid is 1.78×10^{-5} mol dm^{-3}. What is the ratio of conjugate base to conjugate acid in the solution?

Kinetics

Kinetics is the science of measuring changes. In the life and medical sciences, it usually refers to **enzyme kinetics,** the study of the reactions carried out by enzymes.

Enzymes are "biological catalysts" responsible for supporting many of the chemical reactions that maintain homeostasis. Most significant life processes are dependent on enzyme activity.

29.1 Chemical reaction rates

The rate of a chemical reaction is described by the number of molecules of "reactant" that are converted to a "product" in a given time. The reaction rate depends on:

- the concentration of the chemicals involved in the process
- the **rate constant** for that reaction

A reaction in which chemical A (the **reactant**) is converted to B (the **product**) can be written as:

$$A \longrightarrow B$$

The rate of this forward reaction is equal to the product of the molar concentration of A (symbolised by [A]) and the "forward rate constant", k_{+1}

At any one moment, some of chemical B is also being converted back to chemical A:

$$A \longleftarrow B$$

The rate of this reverse reaction is equal to the product of [B] and the "reverse rate constant", k_{-1}

When the rate of the forward reaction is equal to the rate of the reverse reaction, the reaction is said to be in **equilibrium**. When the two chemicals, A and B, are in equilibrium, the ratio of the two concentrations gives the **equilibrium constant** of the reaction, symbolised by K_{eq}. It also equals the ratio of the two rate constants. So,

$$K_{eq} = \frac{[B]}{[A]} = \frac{k_{+1}}{k_{-1}}$$

With a chemical reaction, the higher the concentration of reactant, the faster the reaction, i.e.

Rate of reaction $\propto$ [A]

The $\propto$ symbol means "directly proportional to" (see Section 8.4).

There is no upper limit to the rate of reaction.

29.2 Enzyme kinetics

In reactions catalysed by enzymes, reactants are called "substrates". The substrate and product concentrations are typically thousands of times greater than the enzyme concentration. So, each enzyme molecule will catalyse the conversion of many substrate molecules.

A substrate binds with a specific site, the **active site**, of an enzyme, forming a transitional state called the **enzyme–substrate (ES)** complex. When the ES complex "dissociates" (breaks down), the reaction products are released, and the enzyme is free to link with another substrate molecule. This process can be represented by:

$$E + S \rightarrow ES \rightarrow E + P$$

Some of the ES will dissociate back to E + S, so a more accurate representation is:

$$E + S \rightleftharpoons ES \rightarrow E + P$$

The constants for these three reactions are k_1, k_{-1} and k_2.

$$E + S \underset{k_{-1}}{\overset{k_1}{\rightleftharpoons}} ES \overset{k_2}{\rightarrow} E + P$$

The **Michaelis constant** K_M is defined by:

$$K_M = \frac{(k_{-1} + k_2)}{k_1}$$

If an enzyme has a large K_M it means that it binds to its substrate very weakly – it has a low "affinity" for the substrate, and needs high substrate concentrations to achieve the maximum rate of reaction.

Conversely, an enzyme with a small K_M has a high affinity for its substrate.

29.3 The Michaelis–Menten equation

We have already learnt that, with a *non*-enzymatic reaction, the higher the concentration of reactant, the faster the rate of reaction. There is no upper limit.

However, when an enzyme's catalytic site is working as fast as it can, an increase in the concentration of substrate, [S], will not increase the rate of reaction, V. At this point, the reaction rate is at its maximum, V_{max}.

The **Michaelis–Menten** equation relates the rate of reaction, V, to the substrate concentration, [S], the Michaelis constant K_M, and the maximum rate of reaction, V_{max}, as follows:

$$V = \frac{V_{max}[S]}{K_M + [S]}$$

If the rate of reaction is half of V_{max} the equation becomes:

$$\frac{V_{max}}{2} = \frac{V_{max}[S]}{K_M + [S]}$$

This equation can be manipulated to:

$$K_M = [S]\left\{\left[\frac{2V_{max}}{V_{max}}\right] - 1\right\} = [S]\{2 - 1\} = [S]$$

So, the K_M of an enzyme is the substrate concentration at which the reaction goes at *half* the maximum rate, V_{max}.

Plotting substrate concentration, [S], against rate of reaction, V, using the Michaelis–Menten equation gives the following graph:

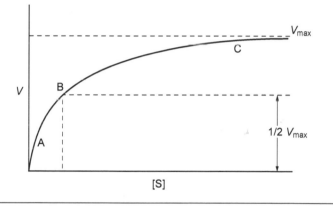

Graph of Michaelis–Menten equation

At point A, when levels of substrate are low, the limiting factor is availability of substrate. The rate of reaction is almost directly proportional to substrate concentration.

As more substrate is added, the rate of reaction increases rapidly.

At point B, $1/2\ V_{max}$, when $[S] = K_M$, 50% of the enzyme sites have substrate bound to them, i.e. are in an ES complex.

Point C, when [S] is high, is close to the point where all of the enzyme molecules have substrate bound to them. The reaction is moving towards its fastest possible rate, and adding more substrate has less and less effect on the rate of the reaction. This is indicated by the graph moving towards its asymptote of V_{max}

However, since V_{max} is approached only slowly as the substrate concentration is increased, it is difficult to calculate an accurate figure for V_{max} from a standard plot such as this. This problem is overcome by using a **Lineweaver–Burk** plot.

29.4 The Lineweaver–Burk plot

The Michaelis–Menten equation, explained above, can be rearranged by taking the reciprocal of both sides of the equation to give:

$$\frac{1}{V} = \frac{1}{V_{max}} + \left(\frac{K_M}{V_{max}} \times \frac{1}{[S]} \right)$$

A plot of $1/V$ against $1/[S]$ is called a **Lineweaver–Burk** plot. It gives a straight line with a gradient of K_M/V_{max}

V_{max} can easily be calculated from the intercept with the y–axis which is at $1/V_{max}$, and K_M can be calculated from the intercept on the x-axis which is at $-1/K_M$

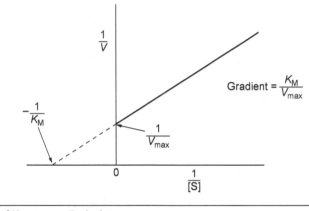

Graph of Lineweaver–Burk plot

Test yourself

The answers are given on page 218.

Question 29.1
An enzyme has a V_{max} of 30 µmol min^{-1} for a particular substrate. Its K_M is 1.5×10^{-5} M. What will the reaction rate be when the substrate concentration is 1.5×10^{-5} M?

Question 29.2
For the same enzyme and substrate, what will the reaction rate be when the substrate concentration is 1.5×10^{-6} M? Give your answer to three significant figures.

Question 29.3
A protease enzyme has a K_M of 1 mM. Plotting experimental findings of the rates of reaction at different substrate concentrations as a Lineweaver–Burk plot gives a straight line with a slope of 0.5. What is the V_{max} of the enzyme?

30 The language of statistics

This chapter explains some of the words that life and medical scientists need to know when analysing data.

30.1 Populations and samples

In statistics, the word **population** means all the individual items which could be studied.

A selection from a population is called a **sample**.

> **EXAMPLE**
>
> A researcher wants to compare broad bean, *Vicia faba*, yield in two fields, one which has been artificially fertilised, the other organically fertilised.
>
> The two populations being studied are all the broad beans in the two fields.
>
> However, she wants to sample a total of 50 m² from each field.

Individuals in the sample may be called **sample units**, **subjects** or **cases**. The data collected from samples are known as **observations**.

The studied differences between individuals in a population are called **variables** or, sometimes, **fields**.

The observations are called **data**.

> **EXAMPLE**
>
> The variables that the researcher wishes to study are the numbers of beans per pod and the mean mass of each sample unit, the individual beans.
>
> The data are the resulting lists of measurements. (Note that the word "data" is a plural, so we say "data are ..." rather than "data is ...")

30.2 Bias

The aim is always that a sample should be representative of a population. If the sample doesn't truly represent the **parent population**, there is said to be **bias**.

An example of this is **observer bias**, where the researcher consciously or subconsciously biases the sample.

To avoid bias, the sample needs to be a **random selection** from the population.

Making a random selection usually involves using **random numbers**. These can be found in random number tables or can be generated by computer programs.

A **quadrat** is an area used as a sample unit.

EXAMPLE

Because there is not time to count and weigh all the beans in both fields, the researcher needs to take samples.

She has a theory that organic farming reduces the numbers of beans per pod and the mass of the beans. She is aware that she could subconsciously choose smaller pods from the organic field, resulting in observer bias.

She also knows that the numbers and masses of the beans may vary in different areas of the same field, so she wants to avoid this sample bias as well.

She therefore tries to avoid these biases by using a random number generator to select five sample areas from each field.

30.3 Variables

Continuous variables are those which can take any value within a given range.

There are two types of continuous variable: ratio and interval.

A **ratio variable** can take any value within a range, i.e. the value needn't be a whole number, but the zero must be a true zero.

EXAMPLE

The length of a bean pod is a ratio variable.

An **interval variable** can take any value within a range, but a value of zero does not indicate an absolute zero.

EXAMPLE

Temperature in °C is an interval variable.

A temperature of 0°C is not a true zero, as the true zero for temperature is −273.15°C (0 K).

A temperature of 10°C is therefore not twice the temperature of 5°C.

Categorical variables are those where only certain values can exist.

Categorical variables can be nominal or ordinal.

A **nominal variable** has no ordering to the categories.

> **EXAMPLE**
>
> Broad bean varieties are nominal variables: we can't have half a variety, and there is no obvious numerical way to put different varieties in order.

An **ordinal variable** has an ordering to the categories.

> **EXAMPLE**
>
> Rows of broad bean may be **ranked** (put in rank order). Ranks can be numbered (1st, 2nd, 3rd rank, etc.). However, the numbers used to label the ranks of ordinal data can't be manipulated mathematically: rank 3 isn't three times rank 1.

A **binary variable** is one that has only two categories, e.g. male and female.

30.4 Descriptive and inferential statistics

Descriptive statistics are those which *describe* the data in a sample. They include means, medians, standard deviations and quartiles. They are designed to give the reader an understanding of the data. They are described in detail in Chapters 31 and 32.

Inferential statistics are statistical methods which make "inferences" about the population, based on the sample of data that has been collected. They can estimate whether the results suggest that there is a real difference in the populations, or how well aspects of a sample are likely to represent the population as a whole. Chapter 35 gives an introduction to inferential statistics.

> **EXAMPLE**
>
> A researcher may give the mean bean yield per square metre in a series of samples. This is a descriptive statistic.
>
> If she then wishes to test whether there is a difference in the mean bean yield per square metre in two fields, she would use inferential statistics.

31 Describing data: measuring averages

When faced with a lot of data, it helps to know what the average is.

There are three ways of giving an average: mean, median and mode. Which of these we need to use depends on the pattern of the data.

31.1 Mean

The **mean** is also known as an arithmetic mean.

It is the commonest measure of a "mid-point", so it's important to have an understanding of how it is calculated.

It is used when the spread of the data is fairly similar on each side of the mid-point, for example when the data are **normally distributed**.

The normal distribution (also known as the "Gaussian distribution") is referred to a lot in statistics. It's the symmetrical, bell-shaped distribution of data shown in the following graph.

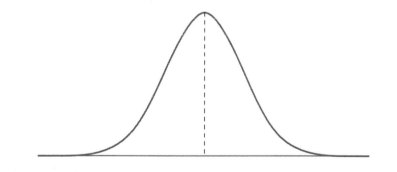

The normal distribution

The dotted line shows the mean of the data.

The mean is the sum of all the values, divided by the number of values.

This can be symbolised by:

$$\bar{x} = \frac{\sum x}{n}$$

where $\bar{x}$ is the symbol for the sample mean (pronounced "x bar"), $\sum x$ is the sum of all the values, and n is the number of values in the sample.

> **EXAMPLE**
>
> Five Scots pines, *Pinus sylvestris*, in a plantation are 3.5, 3.7, 3.8, 3.9 and 4.1 m high.
>
> $$\bar{x} = \frac{\sum x}{n} = \frac{3.5 + 3.7 + 3.8 + 3.9 + 4.1}{5} = \frac{19}{5} = 3.8\,m$$
>
> So the mean height is 3.8 m.

31.2 Population or sample mean?

If we have data on a whole population, then the symbol for the mean is μ, pronounced "mu".

However, if we only have data on a sample, then we use the $\bar{x}$ symbol to symbolise the mean.

> **EXAMPLE**
>
> A researcher wanted to study the mass of all the mallard ducks, *Anas platyrhynchos*, on a lake. He managed to capture and weigh all the ducks. The mean mass was 1.12 kg, so
>
> $$\mu = 1.12\,kg$$
>
> At a second lake he was unable to capture all the ducks, so could only measure a sample of them. In this sample, the mean mass was 1.39 kg, thus
>
> $$\bar{x} = 1.39\,kg$$

31.3 Median

The **median** is the point which has half the values in a sample above, and half below.

If the data are not evenly distributed around their mean, the data are said to be **skewed**. In that case, the mean will not give a good picture of the typical value, so we use the median.

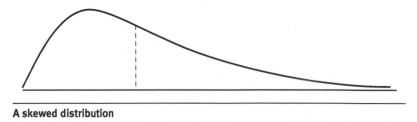

A skewed distribution

The dotted line shows the median.

Compare the shape of the graph with the normal distribution shown in Section 31.1.

EXAMPLE

Let's take the same five Scots pines, *Pinus sylvestris*, as above, measuring 3.5, 3.7, 3.8, 3.9 and 4.1 m.

If there is a 6th pine that is 2.0 m high, then the mean height would be 3.5 m even though only one pine is less than 3.5 m tall. For this skewed sample, the *median* is a more suitable mid-point to use.

Where there are two "middle" values, the convention is that the median is half-way between these.

EXAMPLE

Using the first example of five Scots pines measuring 3.5, 3.7, 3.8, 3.9 and 4.1 m, the median height is 3.8 m, the same as the mean – half the pines are taller, half are shorter.

However, in the second example with six pines measuring 2.0, 3.5, 3.7, 3.8, 3.9 and 4.1 m, there are two "middle" heights, 3.7 m and 3.8 m. The median is half-way between these, i.e. 3.75 m. This gives a better idea of the mid-point of these skewed data than the mean of 3.5 m.

The median may be given with its **quartiles**. The 1st quartile point has $\frac{1}{4}$ of the data below it; the 3rd quartile has $\frac{3}{4}$ of the sample below it. The **inter-quartile range, IQR**, contains the middle $\frac{1}{2}$ of the sample, i.e. the data between the 1st and 3rd quartiles. This can be shown in a **box and whisker** plot.

EXAMPLE

A researcher measured the wing span of a population of 50 small brown bats, *Myotis lucifugus*. The median wing span was 245 mm, 1st quartile 238 mm and 3rd quartile 257 mm. The smallest wingspan was 221 mm, the largest 271 mm. This distribution is represented by the box and whisker plot in the following figure.

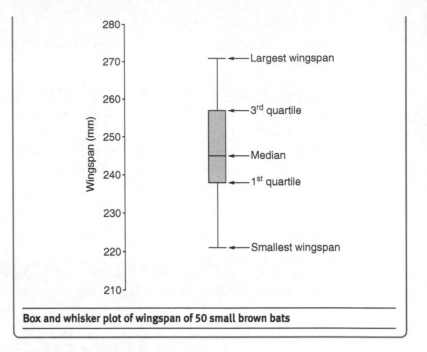

Box and whisker plot of wingspan of 50 small brown bats

The ends of the whiskers represent the maximum and minimum values, excluding any extreme results.

31.4 Mode

The **mode** is the name for the most frequently occurring event.

This is usually only used when we have nominal variables, i.e. those which represent different categories of the same feature and where the categories are not ordered.

EXAMPLE

A researcher noted the eye colour of 100 students. The results are shown here:

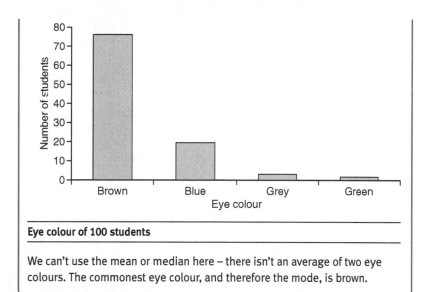

Eye colour of 100 students

We can't use the mean or median here – there isn't an average of two eye colours. The commonest eye colour, and therefore the mode, is brown.

The mode can also be used when there is no one average value, for instance in a **bi-modal** distribution.

EXAMPLE

In this graph there are two "peaks" to the data, i.e. it has a bi-modal distribution.

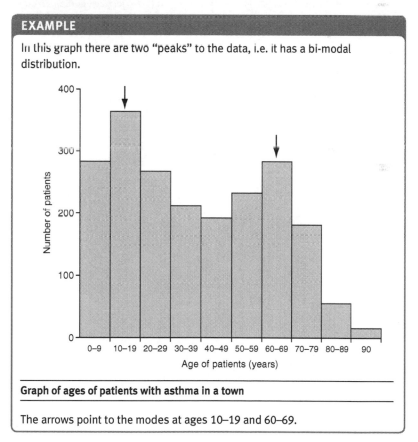

Graph of ages of patients with asthma in a town

The arrows point to the modes at ages 10–19 and 60–69.

Bi-modal data usually suggest that two populations are present that are mixed together, so a mean or median is not a suitable measure for the distribution.

Test yourself

The answers are given on page 218.

Question 31.1
The haemoglobin levels in the blood of five female volunteers are 11.7, 11.9, 12.2, 12.7 and 13.0 g dl^{-1}.
What is the mean haemoglobin level?

Question 31.2
If the haemoglobin levels in the blood of eight female volunteers are 11.1, 11.7, 11.9, 12.3, 12.7, 13.3, 15.2 and 17.4 g dl^{-1}, what is the median?

Question 31.3
The numbers of aphids on 25 lettuce plants are as follows:
44, 45, 37, 32, 54, 44, 52, 59, 40, 39, 46, 50, 32, 49, 32, 41, 42, 43, 53, 53, 46, 43, 56, 45, 42
a) Calculate the mean, median and mode of these values.
b) Which measure of "average" is most appropriate for this sample?

Question 31.4
For each of the following, identify the measure (mean, median, or mode) that provides the best description of the "average" score.
a) A teacher asks a sample of 50 students to name their favourite season (spring, summer, autumn, winter).
b) An insurance company wants to know how long people stay in hospital after a routine operation. The data from a large sample indicate that, while most patients are discharged after 2 or 3 days, some develop infections and stay in the hospital for a few weeks.
c) A researcher uses scores from a standardised reading test administered to a sample of children from a suburban secondary school.

Standard deviation

Standard deviation (SD) is used for data which are normally distributed (see Section 31.1). It provides an indicator of how much the data are spread around their mean.

32.1 Standard deviation

The "deviation" is the difference between any individual reading and the mean of all the readings.

The **standard deviation** is a kind of average of the individual deviations.

So, standard deviation indicates how much a set of values is spread around the mean of those values.

A range of one standard deviation above and below the mean (abbreviated to ±1 SD) includes 68.2% of the values.

±2 SD includes 95.4% of the data.

±3 SD includes 99.7%.

EXAMPLE

Let's say that a group of patients has a normal distribution for weight. The mean weight of the patients is 80 kg. For this group, the SD is calculated to be 5 kg.

- 1 SD below the average is 80 − 5 = 75 kg.
- 1 SD above the average is 80 + 5 = 85 kg.

±1 SD will include 68.2% of the subjects, so 68.2% of patients will weigh between 75 and 85 kg.

95.4% will weigh between 70 and 90 kg (±2 SD).

99.7% of patients will weigh between 65 and 95 kg (±3 SD).

See how this relates to this graph of the data.

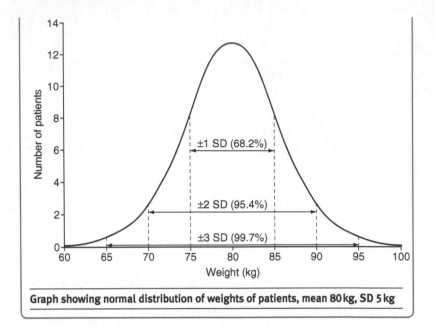

Graph showing normal distribution of weights of patients, mean 80 kg, SD 5 kg

32.2 Calculating the standard deviation of a whole population

Calculating the SD of a *whole population* involves five steps.

1) First we calculate the mean, μ, of the population, N, by adding all the values and dividing the result by the number of values, as in Section 31.1.

2) Then we need to work out the difference (deviation) between each individual value and the mean, by subtracting the mean from each value, symbolised by:

$$x - \mu$$

3) Next we calculate the square of each deviation, multiplying each value by itself. Remember that any negative deviations become positive when squared.

The process so far is:

$$(x - \mu)^2$$

4) Then we calculate the sum of those squared deviations, known as the **sum of squares.**

$$\Sigma (x - \mu)^2$$

5) Then we need the mean of the sum of squares, so we divide the sum by the number of observations. This gives a value called the **population variance, σ^2:**

$$\sigma^2 = \frac{\Sigma (x - \mu)^2}{N}$$

6) The SD is the square root of the variance.

The whole process can be symbolised by:

$$\sigma = \sqrt{\frac{\Sigma (x - \mu)^2}{N}}$$

σ, pronounced "sigma", is the symbol for standard deviation.

EXAMPLE

All 10 salmon, *Salmo salar*, in a lake weigh 1.6, 1.7, 1.8, 1.8, 2.3, 2.4, 2.6, 2.8, 3.1 and 3.3 kg.

NB: normally, whole populations will number far more than 10. We are using a small number so that it is easier to follow the calculations.

1) The sum of the values is 23.4 kg, giving a mean of 2.34 kg.

$$\mu = 2.34 \, kg$$

2) The difference (deviation) between each individual value and the mean is as follows:

1.6 − 2.34	=	−0.74
1.7 − 2.34	−	−0.64
1.8 − 2.34	=	−0.54
1.8 − 2.34	=	−0.54
2.3 − 2.34	=	−0.04
2.4 − 2.34	=	0.06
2.6 − 2.34	=	0.26
2.8 − 2.34	=	0.46
3.1 − 2.34	=	0.76
3.3 − 2.34	=	0.96

3) Calculating the square of each deviation gives:

-0.74^2	=	0.5476
-0.64^2	=	0.4096
-0.54^2	=	0.2916
-0.54^2	=	0.2916
-0.04^2	=	0.0016
0.06^2	=	0.0036
0.26^2	=	0.0676
0.46^2	=	0.2116
0.76^2	=	0.5776
0.96^2	=	0.9216

4) The sum of these squares is 3.324.

5) The mean of the squares is the sum of squares divided by the number of salmon in the population, giving the variance, σ^2:

$$\sigma^2 = \frac{\Sigma\,(x - \mu)^2}{N} = \frac{3.324}{10} = 0.3324$$

6) The square root of the variance gives the SD:

$$\sigma = \sqrt{0.3324} = 0.577\,\text{kg to three significant figures.}$$

32.3 Calculating standard deviation ranges

Having calculated the SD, we can work out the ranges that include 68.2% (± 1 SD), 95.4% (± 2 SD) and 99.7% (± 3 SD) of a population.

EXAMPLE

The mean fasting triglyceride level of 183 patients is found to be 2.2 mmol l^{-1}.

The SD is calculated to be 0.3 mmol l^{-1}.

1 SD below the average is $2.2 - 0.3 = 1.9$ mmol l^{-1}.

1 SD above the average is $2.2 + 0.3 = 2.5$ mmol l^{-1}.

± 1 SD will include 68.2% of the data, so we expect 68.2% of the population to have a triglyceride level between 1.9 and 2.5 mmol l^{-1}.

95.4% will be between 1.6 and 2.8 mmol l^{-1} (± 2 SD).

99.7% of triglyceride levels will be between 1.3 and 3.1 mmol l^{-1} (± 3 SD).

See how this relates to the following graph of the data (99.7% line not shown for clarity).

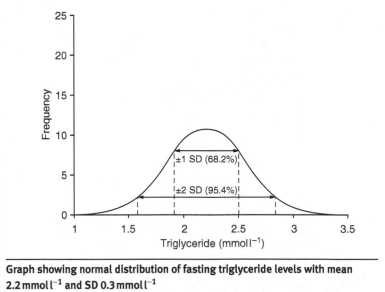

Graph showing normal distribution of fasting triglyceride levels with mean 2.2 mmol l^{-1} and SD 0.3 mmol l^{-1}

32.4 Comparing different standard deviations

If we have two populations with the same mean but different standard deviations, then the population with the larger SD has a wider spread than the population with the smaller SD.

EXAMPLE

If another population of patients has the same mean fasting triglyceride level of $2.2 \, \text{mmol} \, l^{-1}$ but an SD of only $0.2 \, \text{mmol} \, l^{-1}$, ± 1 SD will include 68.2% of the subjects, so 68.2% of the patients will have a fasting triglyceride level of between 2.0 and $2.4 \, \text{mmol} \, l^{-1}$.

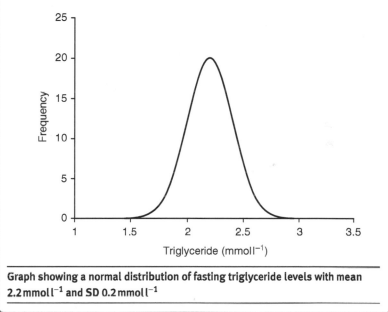

Graph showing a normal distribution of fasting triglyceride levels with mean $2.2 \, \text{mmol} \, l^{-1}$ and SD $0.2 \, \text{mmol} \, l^{-1}$

Compare this graph with the graph of the previous example which is drawn to the same scale. A smaller SD gives a taller, narrower normal distribution.

32.5 Calculating the standard deviation of a sample

When calculating the SD of a *sample*, the variance (denoted as V for a sample but σ^2 for a population) is calculated by dividing the sum of the squared deviations by *one fewer* than the number in the sample, i.e. $n - 1$. The value $n - 1$ is used rather than n as this gives a better estimate of the SD.

EXAMPLE

Let's say that the 10 salmon in the example in Section 32.2 were a *sample* of salmon in a lake. Compare each step with the equivalent calculation above for a population of 10.

1) The sum of the values is still 23.4 kg, giving the same mean of 2.34 kg.

   However, this is now symbolised by $\bar{x} = 2.34$. The use of $\bar{x}$ (rather than μ) denotes that this a sample mean, not the mean of a whole population.

2) The difference (deviation) between each individual value and the mean is also unchanged.

3) Similarly, the calculation for the square of each deviation is unchanged.

4) The sum of squares is also unchanged at 3.324.

5) However, we now divide the sum of squares by (number of salmon − 1) to get the sample variance, V:

$$V = \frac{\sum (x - \bar{x})^2}{n - 1} = \frac{3.324}{9} = 0.3693$$

6) The square root of the variance gives the SD:

$$SD = \sqrt{0.3693} = 0.607 \, kg \text{ to three significant figures.}$$

Test yourself

The answers are given on page 218.

Question 32.1
The haemoglobin levels in the blood of five volunteers are 11.7, 11.9, 12.2, 12.7 and 13.0 gl^{-1}.
What is the standard deviation?

Question 32.2
For the sample of aphids on 25 lettuce plants in *Question 31.3*, calculate the standard deviation for the number of aphids on each plant.

Checking for a normal distribution

Standard deviation should only be used when the data have a normal distribution. However, means and standard deviations are often wrongly used for data which are not normally distributed.

If given only the mean and the SD, a simple check for a normal distribution is to see if 2 SD away from the mean is still within the possible range for the variable.

If we have some data on the mass of a group of men that suggests a mean mass of 75 kg and a standard deviation of 40 kg, then:

$$\text{Mean} - (2 \times \text{SD}) = 75 - (2 \times 40) = -5\,\text{kg}$$

This is clearly an impossible value for the mass of a person, so the data cannot be normally distributed. The mean and standard deviations would therefore not be appropriate measures to use for this sample, as there should be 2.3% of the sample below the 2 SD level.

33.1 z-scores

The number of standard deviation units that an observation is away from the population mean is the **z-score**.

If an observation has a value above the population mean, it has a positive z-score, so

+1 SD gives a z-score of 1.

A value below the mean has a negative z-score, thus

−1 SD gives a z-score of −1.

The formula for the z-score is:

$$z = \frac{(x - \mu)}{\sigma}$$

where z is the z-score, x is the value of the observation, μ is the population mean and σ is the standard deviation.

> **EXAMPLE**
>
> For a population of men, the mean mass is found to be 75 kg, SD 5 kg.
>
> The mass of one man is 65 kg.
>
> $$z = \frac{(x - \mu)}{\sigma} = \frac{(65 - 75)}{5} = -2$$

33.2 A tip

As well as knowing how to calculate the standard deviation, it's worth memorising how much data are included in each standard deviation, so a reminder:

±1 SD includes 68.2% of the data, ±2 SD includes 95.4%, ±3 SD includes 99.7%.

Keeping the "normal distribution" curve in Section 31.1 in mind may help.

Test yourself

The answers are given on page 218.

Question 33.1
The mean blood haemoglobin level in a female population is 12.5 g d l^{-1}, SD 1.2 g d l^{-1}. One woman has a haemoglobin of 15.5 g dl^{-1}. What is the z-score for her haemoglobin?

Question 33.2
A student sits two examinations, getting 57 in the laboratory test and 64 in the written test. The class scores for each examination are normally distributed. For the laboratory test, the mean is 50 with a standard deviation of 6; for the written test, the mean is 50 and the standard deviation is 14.
a) Calculate the z-score for each test.
b) In which of the tests did the student do better compared to the rest of the class?

34 Degrees of freedom

Statisticians use **degrees of freedom**, df, in many of their calculations. The concept is difficult to grasp and, surprisingly, there is no neat definition to explain it.

However, the general concept can be shown as follows.

Let's say that we have three observations (A, B and C) and we want to know their values.

If we know nothing about them other than their existence, then each observation has the freedom of being unknown. There are three degrees of freedom: 3 df.

If we are given the value for the mean, 2.7 say, as soon as we know the values of two of the variables, we can calculate the value of the third. Thus there would be two degrees of freedom: 2 df.

Say we are now told that the standard deviation is 1.2 and the mean 2.7, we can now calculate the values of all the variables as soon as we know the value of any one of them. There is now one degree of freedom: 1 df.

EXAMPLES

If we have two yoghurt pots and we know that one is strawberry and one is chocolate, we don't have to taste both of the yoghurts before we can label both pots. We only need to taste one of them.

$$\text{So, df} = N - 1 = 2 - 1 = 1$$

where N equals the number of yoghurt pots.

If we have 10 yoghurt pots and have a list of their 10 different flavours, we only need to try nine before we can work out the flavour of each pot.

$$\text{Here, df} = N - 1 = 10 - 1 = 9$$

Similarly, if we have a sample of 28 babies and know their mean length, we need to know the length of 27 of them to be able to calculate the length of all of them.

$$\text{df} = N - 1 = 28 - 1 = 27$$

If we also know the standard deviation of their length, then we have again lost a degree of freedom.

$$\text{df} = N - 2 = 28 - 2 = 26$$

How to use statistics to make comparisons

Section 30.4 introduced us to the concept of inferential statistics, those that make "inferences" about a population, based on the sample of data that has been collected.

Inferential statistics estimate whether the results suggest that there is a real difference in the populations, or how well aspects of the sample are likely to represent the population.

This chapter explains the process we need to use when performing inferential statistics.

35.1 The null hypothesis

Inferential statistics involve testing a theory, known as a **hypothesis**.

Paradoxically, for statistical analysis the hypothesis is usually that there is *no* (null) difference between the populations being studied, the **null hypothesis**. The result of the test either supports or rejects that hypothesis.

The null hypothesis is generally the opposite of what we are actually interested in finding out. If we are interested in whether there is a difference between two groups, then the null hypothesis is that there is no difference, and we would try to disprove this.

> **EXAMPLE**
>
> Chapter 30 gave an example of a researcher wanting to compare broad bean, *Vicia faba*, yield in two fields. The null hypothesis is that there is *no* difference between the yields in the two fields based on 5 quadrats from each field.

35.2 Choosing and using the right statistical test

Having collected research data for a comparative study, we need to use a statistical test to compare the data. The **decision-making flowchart** in Appendix 1 will help you decide which test to use.

Applying and calculating the test will give us a **test statistic,** which is a number that quantifies the difference between the samples.

In general, the larger the test statistic result, the larger the difference between the two samples.

> **EXAMPLE**
>
> Using the flowchart in Appendix 1, the researcher decides that the best way to compare the broad bean yield per square metre in the two fields is with the unpaired *t* test. This is described in detail in Chapter 41.
>
> She calculates that the *t* test statistic is 2.51, df 18.

35.3   Significance levels

We now need to work out how likely it is that any difference between the data is due to chance.

We use tables like those in Appendices 2 and 3 to calculate a **significance level** from the test statistic.

If the difference is likely to be due to chance, the difference is said to be "non-significant" and the null hypothesis cannot be rejected.

However, if the difference is not likely to be due to chance, that difference is **significant** and the null hypothesis is rejected.

> **EXAMPLE**
>
> In the table of critical values for *t* distribution in Appendix 2, the *t* test statistic of 2.51 for the bean yield comparison when df is 18 is more than the critical value of 2.10 for 5%, so the difference is unlikely to be due to chance.
>
> If, however, the *t* test statistic for the bean yield comparison has a low significance level, the null hypothesis cannot be rejected – the difference between the two fields could easily be due to chance.

35.4   Are there any confounding factors?

Finally, and outside the scope of this book, if significant differences *are* found, we need to consider whether anything else, any **confounding factors**, could have caused these differences.

> **EXAMPLE**
>
> During her fieldwork, the researcher noticed that the two fields of broad beans differed in various respects: one was on a south-facing slope and was well drained; the other was on level ground and waterlogged.
>
> She realised that these confounding factors could have accounted for the difference in mass per bean, rather than the farming method.

The standard error of the mean

If we take measurements from a random sample of a large population, we can calculate the mean of that sample.

Because of chance, the sample mean is likely to vary from the mean of the whole population (the "population mean", also known as the "true value").

36.1 The standard error of the mean

The **standard error of the mean**, SEM, gives an indication of how close a sample mean might be to the population mean.

The standard error of a sample mean is given by dividing the standard deviation by the square root of the sample size, i.e.

$$SEM = \frac{SD}{\sqrt{n}}$$

The population mean ±1.96 SEM will include 95% of the sample means.

The population mean ±2.58 SEM will include 99% of the sample means.

The population mean ±3.29 SEM will include 99.9% of the sample means.

EXAMPLE

Let's say that we are interested in the mean velocity of fibroblasts at a certain temperature. The "true value" for that mean will be the mean speed of all fibroblasts cultured under those conditions.

We study this by taking a sample of 10 fibroblasts and measuring their mean velocity. Because of random variation, the mean of this sample is likely to vary from the "true value". We don't know for certain how close it is to the true value, but the larger the sample, the closer it is likely to be.

Another sample would be likely to give slightly different results. If we took 10 separate samples, each would give a different result. However, we can see that the means would be distributed around a central area – this area would probably contain the true population mean.

This distribution of means around the population mean will be a normal distribution and can be described by the standard error of the mean.

If the sample sizes are larger (30 fibroblasts, say, instead of 10), then the standard error will be smaller, and the distribution will be narrower.

36.2 How SEM works

Even with only one sample, we know that the sample mean is part of that distribution around the population mean, we just don't know where it is. The standard error of the mean is the standard error of that distribution. It takes into account the sample size and the variation within that sample.

> **EXAMPLE**
>
> In Section 32.5 we calculated the mean mass of 10 salmon to be 2.34 kg. The standard deviation was 0.607 kg.
>
> $$SEM = \frac{SD}{\sqrt{n}} = \frac{0.607}{\sqrt{10}} = 0.1929 \, kg$$

36.3 The effect of a larger sample size

Increasing sample size reduces the standard error.

> **EXAMPLE**
>
> If we quadruple the sample size of salmon and the standard deviation is still 0.607 kg, then
>
> $$SEM = \frac{SD}{\sqrt{n}} = \frac{0.607}{\sqrt{40}} = 0.0960 \, kg$$
>
> So, quadrupling the sample size halves the standard error.

36.4 Standard deviation or standard error?

Standard deviation tells us how much the data in *samples* vary around their own mean.

We use standard error when we want to know how much the sample *means* vary around their population mean.

Test yourself

The answers are given on pages 218–19.

Question 36.1
The mean blood haemoglobin level in 16 pregnant women is 11.6 g dl^{-1}, SD 0.4 g dl^{-1}. What is the SEM?

Question 36.2
The number of samples in *Question 36.1* is increased to 36, but the mean and SD remain the same. What is the SEM now?

Confidence intervals

If we have the mean value of a sample and want the range that is likely to contain the true population value, we can use the standard error to calculate the **confidence interval**, CI.

The confidence interval is the range (interval) in which we can be fairly sure (confident) that the population mean lies, i.e. the mean value that we would get if we had data for the whole population that we have sampled.

37.1 The 95% confidence interval

The population mean ±1.96 standard errors will include 95% of the sample means.

It follows that, for any given sample, there is a 95% chance (we can be 95% confident) that the sample mean is within ±1.96 standard errors of the population mean.

This is often taken to mean that there is a 95% chance that the population mean is within ±1.96 standard errors of the sample mean. Although technically not correct, the concept is commonly used.

37.2 The effect of sample size and SD on the confidence interval

The size of a confidence interval is related to the sample size and the size of the standard deviation:

- larger studies are likely to have narrower confidence intervals;
- the smaller the SD, the narrower the confidence interval.

37.3 Calculating the 95% confidence interval

The range from 1.96 standard errors below the sample mean to 1.96 standard errors above it is called the 95% confidence interval.

To calculate the 95% confidence interval, we first need to multiply the standard error by 1.96.

Subtracting the result from the sample mean gives the lower 95% **confidence limit**; adding it to the sample mean gives the upper 95% confidence limit.

$$95\% \text{ CI} = \bar{x} \pm (\text{SEM} \times 1.96)$$

> **EXAMPLE**
>
> In Section 36.2 we found that the mean mass of a sample of 10 salmon was 2.34 kg, with a standard error of the mean of 0.1929 kg.
>
> To calculate the 95% confidence interval:
>
> $$\text{SEM} \times 1.96 = 0.1929 \times 1.96 = 0.3781$$
>
> $$\bar{x} - 0.3781 = 2.34 - 0.3781 = 1.9619$$
>
> $$\bar{x} + 0.3781 = 2.34 + 0.3781 = 2.7181$$
>
> As stated above, while technically incorrect, this is usually interpreted as being 95% confident that the true population mean for this sample of fish is between 1.96 and 2.72 kg, to three significant figures.
>
> This is written as:
>
> mean mass 2.34 kg, 95% CI 1.96–2.72 kg.

37.4 Calculating other confidence intervals

The sample mean ±2.58 SEM gives the 99% confidence interval.

The sample mean ±3.29 SEM gives the 99.9% confidence interval.

> **EXAMPLE**
>
> We wish to calculate the 99% confidence interval for the salmon in the example above.
>
> $$\text{SEM} \times 2.58 = 0.1929 \times 2.58 = 0.4977$$
>
> $$\bar{x} - 0.4977 = 2.34 - 0.4977 = 1.8423$$
>
> $$\bar{x} + 0.4977 = 2.34 + 0.4977 = 2.8377$$
>
> This is usually interpreted as being 99% confident that the true population mean for these fish is between 1.84 and 2.84 kg.
>
> Similarly, to calculate the 99.9% confidence interval:
>
> $$\text{SEM} \times 3.29 = 0.1929 \times 3.29 = 0.6346$$
>
> $$\bar{x} - 0.6346 = 2.34 - 0.6346 = 1.7054$$
>
> $$\bar{x} + 0.6346 = 2.34 + 0.6346 = 2.9746$$
>
> This is usually interpreted as being 99.9% confident that the true population mean for the salmon is between 1.71 and 2.97 kg.

Test yourself

The answer is given on page 219.

Question 37.1
The mean blood haemoglobin level in a sample of 20 women is $12.8\,g\,dl^{-1}$. The SEM is $0.8\,g\,dl^{-1}$.
a) Give the 95%, 99% and 99.9% confidence intervals.
b) Calculate the SD to three significant figures.
c) If the sample has increased to 200 women, but the mean and SD are still the same, what are the 95%, 99% and 99.9% confidence intervals?

Probability

An understanding of **probability** is key to many statistical analyses.

The concept of chance is something that we can intuitively understand. The many ways of describing probability can be confusing, however.

38.1 Five possible ways of stating a probability

A numerical value for a probability can be given in different ways.

> **EXAMPLE**
>
> If we toss a coin, there is an equal chance that it will land heads or tails.
>
> So, there is a 1 in 2 probability that it will land heads up.

Probability can be written as a fraction. We divide the number of nominated outcomes by the total number of possible outcomes.

> **EXAMPLE**
>
> With a coin, the probability of throwing heads is 1 divided by 2: we are likely to get heads 1/2 the number of times that we toss the coin.

Probability is expressed on a scale from 0 to 1:

- a rare event has a probability close to 0;
- a very common event has a probability close to 1.

> **EXAMPLE**
>
> Probability of throwing a head $= \dfrac{1}{2} = 0.5$

Probability can also be written using the abbreviation "P" or "p"

> **EXAMPLE**
>
> For a coin, the chance of throwing a head is $P = 0.5$

Additionally we can state chance as a percentage.

> **EXAMPLE**
>
> When tossing a coin, there is a 50% chance of throwing a head.

So, 1 in 2, 1/2, 0.5, P = 0.5 and 50% all give the same information about the chance of a coin landing heads up.

EXAMPLE

The 95% confidence interval for the size of a strain of *E. coli* bacteria is 1.9 to 2.1 μm.

This is usually interpreted as meaning that we are 95% confident that the true population mean is between 1.9 and 2.1 – and there is a 5% chance that it isn't.

We can also describe the 95% probability as

- a 19 in 20 chance
- a 19/20 chance
- a 0.95 probability
- P = 0.95

Test yourself

The answers are given on page 219.

Question 38.1
Use five different ways to describe the chance of throwing two sixes with one roll of a pair of dice.

Question 38.2
What is the chance of getting all sixes in a single throw of six fair dice?

39 Significance and P values

Significance is an important concept related to probability.

When comparing measurements of two or more groups, we may make a hypothesis on the difference between them. The P (probability) value is used when we wish to see how likely it is that the hypothesis is true.

As described in Section 35.1, the hypothesis is usually that there is *no* difference between the groups, known as the "null hypothesis".

The significance level describes the likelihood that the null hypothesis is correct.

39.1 What significance means

If we take two groups of patients who have been given different treatments, and the resulting mean cure rates are different, we may want to know whether there is a significant difference between the two means. Has the difference happened by chance, or might there truly be a difference between the two groups?

The hypothesis is that there is *no* difference between treatments, known as the **null hypothesis**.

> **EXAMPLE**
>
> Two hundred adult patients with bronchopneumonia have been randomised to receive one of two possible antibiotics. Five days later, they are reassessed.
>
> The researchers wish to know how likely it is that any difference between the effects of the two treatments could have happened by chance, or whether there is a significant difference.
>
> The null hypothesis is that there is no difference between the effects of the two treatments.

39.2 The P value

The **P value** gives the probability of any observed difference in measurement between the two groups having happened by chance.

P = 0.5 means that the probability of the difference having happened by chance is 0.5 in 1.

P = 0.05 means that the probability of the difference having happened by chance is 0.05 in 1, i.e. 1 in 20. It is the figure frequently quoted as being "statistically significant", i.e. unlikely to have happened by chance and therefore important. However, this is an arbitrary figure. If we look at 20 studies, even if there is no difference in any of the groups studied, one of the studies is likely to have a P value of 0.05 and so appear significant!

The lower the P value, the less likely it is that the difference happened by chance, and so the higher the significance of the finding.

P = 0.01 is often considered to be "highly significant". It means that the difference will only have happened by chance 1 in 100 times. This is unlikely, but still possible.

P = 0.001 means the difference will have happened by chance 1 in 1000 times; even less likely, but still just possible. It's usually considered to be "very highly significant".

EXAMPLE

An epidemiologist finds that in one town, out of 50 babies, 35 are female.

She wants to know the probability that this difference from the usual 50:50 male:female ratio in the rest of the county happened by chance.

The null hypothesis is that in this area the chance of having a female baby *hasn't* been altered.

The P value gives the probability of obtaining the observed, or more extreme, results assuming the null hypothesis is true.

The P value in this example is 0.007. Other sections show how this is calculated, but at the moment just concentrate on what it means.

P = 0.007 means the result would only have happened by chance in 0.007 in 1 (or 1 in 140) times if the choice of the area didn't actually affect the sex of the babies. This is highly unlikely, i.e. "highly significant", so we can reject our hypothesis and conclude that there is a highly significant difference between the sex ratio in this town and the rest of the county.

39.3 Does statistical significance always equal relevance?

Try not to confuse statistical significance with relevance. If a sample is too small, the results are unlikely to be statistically significant even if there really is a difference between them. Conversely a large sample may find a statistically significant difference that is too small to have any relevance.

Test yourself

The answers are given on page 219.

Question 39.1
We wish to study the effect of two different temperatures on the germination rate of common wheat seeds, *Triticum aestivum*. Set up a null hypothesis to test this.

Question 39.2
The germination rate of a sample of wheat seeds at 10°C is found to be 92%. The germination of another sample of wheat seeds, kept at 14°C, is 96%; P = 0.25.
How significant is this difference?

Tests of significance

There is a large array of significance tests, and it is not always easy to know which should be used when.

The **decision-making flowchart** in Appendix 1 is designed to help you decide which test to use.

There are two main groups of statistical tests: **parametric** and **non-parametric**. The choice of which to use depends on the distribution of the data.

40.1 Parametric tests

Generally, parametric tests compare means and variances. They should be used only when the data follow a **normal distribution**, the bell-shaped curve shown in Section 31.1.

For large samples (above 50, say) the sample *means* will usually be normally distributed even if the samples themselves aren't, so it may be possible to use parametric tests.

Some skewed data can be **transformed** to normally distributed data, which can then be analysed using the more accurate parametric testing. For instance, a skewed distribution might become normally distributed if the logarithms of the values are used.

You may see reference to the **Kolmogorov Smirnov** test. This tests the hypothesis that data are from a normal distribution, and therefore assesses whether parametric statistics can be used.

Statisticians prefer to use parametric tests when possible because:

- for parametric data, parametric tests are more powerful than non-parametric tests, and
- there are far more parametric tests readily available.

However, sample populations that don't meet (or cannot be transformed to meet) parametric criteria need to be analysed with non-parametric tests.

40.2 The commonly used parametric tests

t **tests** are used to compare sample means. They are described in detail in Chapter 41.

> **EXAMPLE**
>
> We have two sets of 3-month-old pigeons that have been fed on different grain. We can use a *t* test to compare their mean masses.

When looking for an association between two categorical variables we analyse the data using the **chi-squared** test (see Chapter 43).

> **EXAMPLE**
>
> If we studied the possible effect of a pollutant on the severity of asthma, we could label asthma severity as mild, moderate or severe. The effect of the presence or absence of the pollutant on severity would be analysed with the chi-squared test.

We can test the hypothesis that samples come from the same population by looking at the variation *within* each sample, and comparing that with the variance (see Section 32.2) *between* the sample means. This is called **analysis of variance** (ANOVA), and is particularly useful when comparing multiple variables. It is discussed in Chapter 42.

> **EXAMPLE**
>
> A study of the effect of five different fertilisers on a variety of crops needs analysis of variance.

Correlation studies the strength of a linear (straight-line) relationship between two variables. This is explained in Chapter 44.

> **EXAMPLE**
>
> If we wish to know the strength of a link between the incidence of obesity and diabetes in different age groups, we use Pearson's correlation test.

Regression analysis, explained in Chapter 45, quantifies how one set of data relates to another when one of the variables is dependent on the other, independent, variable.

> **EXAMPLE**
>
> The relationship of hours of daylight (an independent variable) to plant growth (the dependent variable) can be quantified with regression analysis.

40.3 Non-parametric tests

When data cannot satisfy (or be transformed to satisfy) the requirements for analysis using a parametric test, we need to use a non-parametric test.

In general, non-parametric tests compare medians (see Section 31.3).

Rather than comparing the values of the raw data, the tests put the data in **ranks** and compare the ranks.

The non-parametric equivalents to parametric tests are given in the following table.

Table of parametric tests and their non-parametric equivalents	
Parametric test	**Non-parametric equivalent**
Mean	Median or mode
Standard deviation	Quartiles and interquartile range
One-sample t test	Wilcoxon test, sign test
Paired t test	Wilcoxon test, sign test
Unpaired t test	Mann–Whitney U test
One-way ANOVA	Kruskal–Wallis test or ANOVA on ranked data
Repeated measures ANOVA	Friedman test or ANOVA on ranked data
Pearson's correlation test	Spearman's rank correlation coefficient

A detailed description of the non-parametric tests is outside the scope of this book.

40.4 Which to use when

We cannot make comparisons between samples until we have decided whether to use parametric or non-parametric tests. However, the decision on which to use is surprisingly controversial – different statisticians can give different advice.

A simplified decision-making chart is given here, but you may get other equally valid advice.

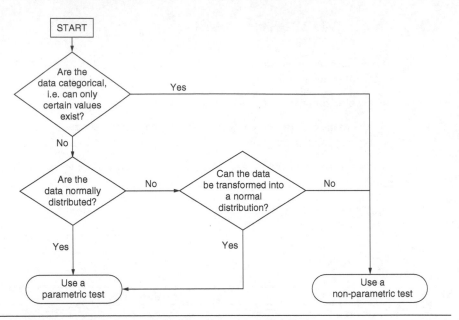

Flowchart for choosing between parametric and non-parametric tests

Test yourself

The answers are given on page 219.

Question 40.1
Forty patients have completed a quality of life questionnaire. The questionnaire consisted of basic demographic information as well as 20 questions with five-point rating scales. Use the flow chart on this page to decide whether a parametric or non-parametric test is needed for each of the following areas of interest:
a) Whether there is a significant age difference between the men and women in the sample.
b) One of the five-point rating scale questions relates to the quality of life in the respondents' home environments. You want to know whether there is a significant difference between the answers of the men and the women.
c) An overall quality of life score can be calculated by adding together the results of all 20 five-point rating scale questions. Again, you would like to compare men's and women's scores. Previous researchers have found the overall quality of life score to be normally distributed.

41 *t* tests

Like other parametric tests, the ***t* test** (correctly termed **Student's *t* test**) is used to compare samples of normally distributed data (see Section 31.1) with similar standard deviations. *t* tests are typically used to compare one or two samples. They test the probability that the samples come from a single population with a single mean value.

For small samples, the *z*-score (see Section 33.1) does not provide a good estimate of the distribution of differences between groups. However, the *t* score has been developed to provide a better estimate that overcomes this.

41.1 *t* test tables

The table of *t* values in Appendix 2 gives us the significance level of the difference between two means for a given sample size and *t* score.

We reject the null hypothesis if the calculated value of *t* is larger than the value on the table for a chosen significance level.

> **EXAMPLE**
>
> We wish to compare the masses of two samples of 10 eggs from different chicken species. The null hypothesis is that there is no difference between the masses.
>
> A *t* test on the two samples gives a *t* score of 2.62, written as $t = 2.62$.
>
> The two samples of 10 eggs means 18 degrees of freedom, df (see Chapter 34). Using the table in Appendix 2, we can see that for 18 df, at the 5% significance level, $t = 2.10$.
>
> Our *t* score of 2.62 is larger than this value, so the significance level is less than 5%, i.e. $P < 0.05$.
>
> We can be sure that the chance of getting results as extreme as these when there is actually no difference between the species is less than 5%, so can reject the null hypothesis.

41.2 One-tailed and two-tailed tests

Under a normal curve 95% of observations are within 1.96 standard deviations of the mean.

The remaining 5% are equally divided between the **tails** of the normal distribution, as shown below.

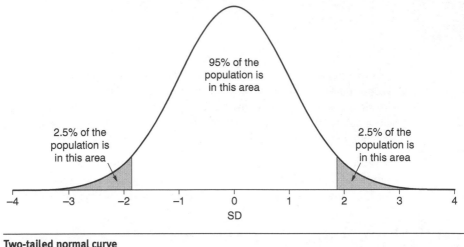

95% of the population is in this area

2.5% of the population is in this area

2.5% of the population is in this area

SD

Two-tailed normal curve

When trying to reject a null hypothesis (Section 35.1), we are generally interested in two possibilities: either we can reject it because the mean value of one sample is higher than that of the other sample, or because it is lower.

By allowing the null hypothesis to be rejected from either direction we are performing a **two-tailed test** – we are rejecting it when the result is in either tail of the test distribution.

However, if we know that a measurement in a population is larger (or smaller) than another, then we know that the residual 5% will be in the upper (or lower) tail of the normal distribution, as shown below:

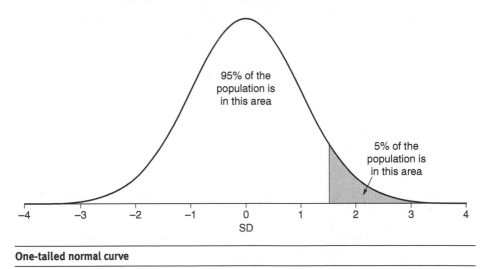

95% of the population is in this area

5% of the population is in this area

SD

One-tailed normal curve

In this case, the critical values for *t* tests will be lower, and we use a **one-tailed test**, rejecting the null hypothesis only when the result is in a single tail of the test distribution.

Statistical software can calculate for both one- and two-tailed tests, and tables like the table of critical values for *t* distribution in Appendix 2 give significance levels for both.

In practice, usually there is the possibility of both improvement and deterioration, and therefore rarely the need to use one-tailed tests.

Also, a P value that is not quite significant on a two-tailed test may become significant if a one-tailed test is used. Researchers have been known to use this to their advantage!

41.3 Three different *t* tests

There are three different *t* tests.

When we have a single sample and wish to compare its mean with a fixed value, for example the population mean, we use the **one-sample *t* test**.

Where two observations are made on the same sample subjects, we need the **paired *t* test**. (Confusingly, it is also known as the related *t* test and paired-samples *t* test.)

Where we have two samples and have measured the same variable on each of them, the **unpaired *t* test** (also known as the two-sample *t* test and independent-samples *t* test) is used.

EXAMPLES

We are interested in the mean mass of day-old chicken eggs at a farm.

We have a sample of 10 eggs and wish to know whether their mean mass is significantly different to a national standard. For this, we need the one-sample *t* test.

If we weigh the *same* sample of eggs a day later and wish to know whether their mass has changed significantly, we use the paired *t* test.

If, however, we weigh a *different* sample of day-old eggs and want to know whether their mass is significantly different, we need the unpaired *t* test.

41.4 The one-sample *t* test

The one-sample *t* test compares the mean of a single sample with a fixed value, for example the population mean.

The *t* value is the number of standard errors that the sample mean is away from the population mean.

The formula is:

$$t = \frac{\bar{x} - E}{\text{SEM}}$$

where $\bar{x}$ is the sample mean, E is the fixed value, and SEM is the standard error of the mean for the sample (how to calculate these values was explained in Chapters 31 and 36).

EXAMPLE

Using the example above, let's say that the mean mass of our sample of 10 eggs is 60 g, and the SEM of the mean is 1.6 g.

A national standard mass for day-old eggs is 55 g.

$$t = \frac{\bar{x} - E}{\text{SEM}} = \frac{60 - 55}{1.6} = 3.125$$

In the table of t values in Appendix 2, for 9 df a value of $t = 3.125$ is more than the 2.26 needed for the 5% significance level.

Thus the difference between the observed mean and the national standard is unlikely to have happened by chance.

41.5   The paired t test

The paired t test compares the means of a variable in the same sample under different conditions, or at two different times.

It is calculated by dividing the mean of the differences between each pair by the standard error of the difference.

Its formula is:

$$t = \frac{\bar{d}}{\overline{\text{SE}}_d}$$

where $\bar{d}$ is the mean difference between each pair and $\overline{\text{SE}}_d$ is the standard error of the differences.

The paired samples t test is equivalent to a one-sample t test on the difference between the two samples and giving E a value of zero.

EXAMPLE

Each of the eggs in our sample of 10 is weighed on day 1 and again the next day. We wish to know whether there is a significant difference between these two masses.

The standard error of their differences is calculated to be 1.05 g. The mean change of mass is 2 g.

$$t = \frac{\overline{d}}{\overline{SE_d}} = \frac{2}{1.05} = 1.905$$

Using the table of *t* values in Appendix 2, for 9 df the critical value of *t* for 5% is 2.26. Our calculated value of 1.905 is therefore less than needed for the 5% significance level, so the difference may well have happened by chance.

41.6 The unpaired *t* test

The unpaired *t* test compares the means of the same variable in two different samples.

It is calculated by dividing the difference between the means of the two samples by the standard error of the differences.

The formula is:

$$t = \frac{\overline{x_a} - \overline{x_b}}{\overline{SE_d}}$$

where $\overline{x_a}$ and $\overline{x_b}$ are the means of the two samples, and $\overline{SE_d}$ is the standard error of the differences.

EXAMPLE

Our first sample of day-old eggs has a mean mass of 60 g. A different sample of day-old eggs has a mean mass of 51 g. We want to know if there is a significant difference in their masses.

The standard error of their differences is calculated to be 2.24 g.

$$t = \frac{\overline{x_a} - \overline{x_b}}{\overline{SE_d}} = \frac{60 - 51}{2.24} = 4.018$$

From the table of *t* values in Appendix 2, we can see that for 18 df a value of $t = 4.018$ is more than the 3.92 needed for the 0.1% significance level, so the difference is very unlikely to have happened by chance.

Test yourself

The answers are given on page 219.

Question 41.1
The germination rate for foxglove seeds, *Digitalis* spp., produced by a large nursery is 70%.
Twelve samples of seeds have been stored for 1 year. The mean germination rate of the samples is 62%, SEM 8%.
What is the *t* value? Use the table on *t* values in Appendix 2 to work out how likely it is that the difference is due to chance.

Question 41.2
The mean core temperature of a group of 20 volunteers is 36.80°C when the ambient temperature is 30.0°C.
The mean increase in their core temperature is 0.1°C after exposure to an ambient temperature of 40.0°C.

The standard error of their differences is calculated to be 0.04°C.
What is the *t* value? Use the table in Appendix 2 to work out how likely it is that the difference is due to chance.

Question 41.3
The mean height of a sample of 30 common sunflower plants, *Helianthus annuus*, is 1.5 m.
Another sample has been grown in soil of lower humidity, and has a mean height of 1.2 m.
The standard error of their difference is 0.1 m.
What is the *t* value? Use the table on *t* values in Appendix 2 to work out how likely it is that the difference is due to chance.

Analysis of variance

Analysis of variance, also known by its acronym ANOVA, is used to compare multiple samples.

It is a powerful statistical technique that can be used to compare, for example, multiple samples, from multiple groups, under the influence of multiple factors.

42.1 ANOVA or *t* test?

Where we want to compare two means we use the *t* test.

Comparing three or more means could be done by multiple use of the *t* test. Comparing the sample means $\bar{A}$, $\bar{B}$ and $\bar{C}$, for example, could be done by using the *t* test three times: comparing $\bar{A}$ with $\bar{B}$, $\bar{A}$ with $\bar{C}$, and $\bar{B}$ with $\bar{C}$.

The larger the number of samples, the more *t* tests we would need to do.

ANOVA has the advantage that a single test covers all the comparisons, and it should therefore be used when comparing more than two samples.

ANOVA also confers the advantage of being able to consider the effect of multiple factors on a variable of interest.

> **EXAMPLE**
>
> If we wish to test the null hypothesis that there is no difference in the yield of seven varieties of tomato, we would need to do 21 *t* tests.
>
> A single ANOVA analysis would take the place of the 21 *t* tests.
>
> We could also use ANOVA to consider the effect of both variety and field position on the yield of the crop.

42.2 Problems with multiple testing

When a *t* test gives a P value of 0.05, there is still a 5% possibility that we should not have rejected the null hypothesis and therefore a 5% chance that we have come to the wrong conclusion.

If we do lots of independent *t* tests, then this chance of making a mistake will be present each time we do a test. Therefore, the more tests we do, the greater the chances of drawing the wrong conclusion.

Because ANOVA is a single test, the problems of multiple testing don't apply.

42.3   How ANOVA works

Calculating ANOVA is, however, a complex process, so we shall simply give an outline of how it works.

ANOVA compares the variability *between* samples with the variability *within* samples.

EXAMPLE

This plot shows the data from two samples.

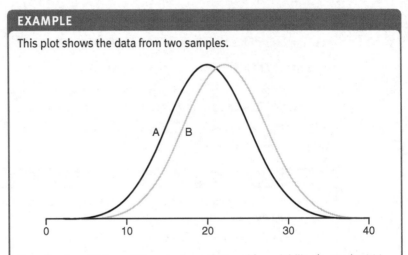

Samples A and B have different means, but a wide variability (scatter) within them. They could have come from the same population.

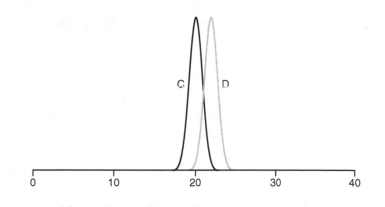

However, while samples C and D have the same means as in the previous plot, they each have a low variability, so probably come from different populations.

One-way ANOVA is used when we wish to compare means from more than two samples. It may therefore be considered as an extension of the t test.

Repeated measures ANOVA is used when there are repeated measurements on the same sampling unit.

42.4 The *F* value

The ANOVA test statistic, F, is derived from the ratio of the mean between-sample variations and the mean within-sample variations.

$$F = \frac{\text{between-sample variance}}{\text{within-sample variance}}$$

Calculating the F value is complex and best left to statistical software. The software will also calculate the P value for you.

EXAMPLE

A researcher is interested in assessing whether three different samples of patients on different lipid-lowering medication have significantly different fasting cholesterol levels. Analysing the results using statistical software gives the following results.

Table of ANOVA results

	Sum of squares	df	Mean square	F	Sig.
Between groups	2604.205	2	1302.102	1.761	0.192
Within groups	19228.002	26	739.539		
Total	21832.207	28			

The higher the F value, the more likely it is that the difference between the samples is statistically different. In this example we can see that we have an F value of 1.761 on two degrees of freedom. From the table we can see that the significance level (Sig.) is 0.192, and therefore there is a 19.2% possibility that the observed differences between the samples have arisen by chance.

42.5 Finding out which samples are different

ANOVA tells us whether there is a significant difference between samples, but it doesn't tell us *which* of those samples have that difference.

For this we need further tests, known as **post hoc tests**. Again, post hoc tests such as the **Bonferroni correction** and **Dunnett's**, **Scheffe** and **Tukey** tests are available on statistical software.

Test yourself

The answer is given on page 219.

Question 42.1
A clinician has studied lung function in a group of 34 patients. She has produced the following ANOVA tables that analyse the impact of their smoking status and gender on their forced expiratory volume in 1 second (FEV_1).

Between-subject factors

		Value label	N
Gender	F		16
	M		18
Smoker	2	No	7
	3	Ex-smoker	26
	4	Passive	1

Tests of between-subject effects

Dependent variable: FEV_1

Source	Type III sum of squares	df	Mean square	F	Sig.
Corrected model	0.873*	3	0.291	2.247	0.103
Intercept	13.072	1	13.072	101.005	0.000
Gender	0.603	1	0.603	4.663	0.039
Smoker	0.246	2	0.123	0.952	0.397
Error	3.883	30	0.129		
Total	59.340	34			
Corrected total	4.755	33			

*$R^2 = 0.183$.

a) Comment on the significance of gender and smoking status as predictors of FEV_1.
b) How good is this model in accounting for the variation in FEV_1?

The chi-squared test

The **frequency** of an event is the number of times that it occurs.

Chi-squared is a measure of the difference between actual and expected frequencies.

Usually written as χ^2, chi is pronounced as in "sky" without the s.

43.1 Expected frequency

The **expected frequency** is the frequency if there is *no* difference between sets of results (the null hypothesis).

We can use a **contingency table** to compare expected and actual frequencies.

> **EXAMPLE**
>
> In one area of forest, 15 out of 31 chimpanzees (Sample A) are found to be male. In another area, 36 out of 60 are male (Sample B). We wish to know whether this is a statistically significant difference.
>
> | Contingency table for sex of chimpanzee offspring | | | |
> |---|---|---|---|
> | | Sample A | Sample B | Totals |
> | Males | 15 | 36 | 51 |
> | Females | 16 | 24 | 40 |
> | Totals | 31 | 60 | 91 |

43.2 Calculating chi-squared

This is given by:

$$\chi^2 = \Sigma \frac{(O - E)^2}{E}$$

where $(O - E)$ is the difference between observed and expected frequencies, and E is the expected frequency; the Σ symbol means "sum of".

EXAMPLE

In the chimpanzee example, the expected frequency, E, of males in sample A is given by:

$$E = \frac{\text{Total number of males} \times \text{total number in sample A}}{\text{Total number of chimpanzees}}$$

So, $E = 51 \times 31/91 = 17.374$ for males in Sample A; $E = 40 \times 31/91 = 13.626$ for females in Sample A; $E = 51 \times 60/91 = 33.626$ for males in Sample B; $E = 40 \times 60/91 = 26.374$ for females in Sample B.

To calculate $\sum \dfrac{(O - E)^2}{E}$ we need to:

- subtract the expected from the observed frequencies,
- take the square of each of these differences,
- divide each squared difference by the expected values,
- take the sum of the results.

We have shown the calculations in the table below:

Table calculating χ^2 for chimpanzee example					
	Observed frequency, O	Expected frequency, E	$O - E$	$(O - E)^2$	$\dfrac{(O - E)^2}{E}$
Males in Sample A	15	17.374	−2.374	5.636	0.3244
Females in Sample A	16	13.626	2.374	5.636	0.4136
Males in Sample B	36	33.626	2.374	5.636	0.1676
Females in Sample B	24	26.374	−2.374	5.636	0.2137
				$\sum \dfrac{(O - E)^2}{E} =$	1.1193

So $\chi^2 = 1.12$, to three significant figures.

43.3 Calculating the significance level

Once we know the χ^2 value, we can work out the significance level.

The significance level for χ^2 depends on the number of degrees of freedom, df, explained in Chapter 34.

The degrees of freedom for this test is the number of rows in the table less 1, multiplied by the number of columns in the table less 1, i.e.

$$df = (\text{rows} - 1)(\text{columns} - 1)$$

Critical values for χ^2 (usually written as X^2) are given in Appendix 3.

EXAMPLE

For the chimpanzee example above, there are two rows (males and females) and two columns (samples A and B), so:

$$df = (2 - 1)(2 - 1) = 1$$

The table in Appendix 3 shows that the critical value for the 5% significance level is 3.84 for 1 df.

But, in our example, χ^2 is only 1.12, so the result is not significant and the null hypothesis stands.

43.4 Other tests for contingency tables

Instead of the χ^2 test, **Fisher's exact test** can be used to analyse contingency tables. Fisher's test is the best choice for 2 by 2 tables, with expected frequencies of fewer than 5.

The χ^2 test is simpler to calculate but gives only an approximate P value and is inappropriate for small samples. We can apply **Yates' continuity correction**, or other adjustments, to the χ^2 test to improve the accuracy of the P value.

The **Mantel–Haenszel** test is an extension of the χ^2 test that is used to compare several two-way tables.

Test yourself

The answers are given on pages 219–20.

Question 43.1
A bacteriologist inoculates 480 Petri dishes with *E. coli* bacteria; 240 of the Petri dishes contained a standard culture medium, 240 had a new type of culture medium. After 3 days she checks whether or not there are any bacterial colonies. The results are shown below.

Table of effects of type of culture medium on *E. coli* growth			
	Type of culture medium		
	Standard	New	Total
Bacterial colony present at 3 days	144	160	304
No bacterial colony at 3 days	96	80	176
Total	240	240	480

Calculate the χ^2 value. You will find a calculator helpful.

Question 43.2
For your calculated χ^2 value for the *E. coli* example, use the table in Appendix 3 to work out whether it is significant.

Question 43.3

A population of fungal spores was categorised
into two sizes, large and small. The spores were
incubated on agar and the numbers of spores
that germinated by producing either a single
outgrowth or multiple outgrowths were counted.

	Large spores	Small spores	Total
Multiple outgrowth	80	18	98
Single outgrowth	40	42	82
Total	120	60	180

Does this provide evidence of a significant
relationship between the size of spores and
whether the outgrowth is single or multiple?

Correlation

Where there is a linear relationship between two variables there is said to be a **correlation** between them.

44.1 Positive or negative?

A **positive** correlation coefficient means that as one variable is increasing, the value for the other variable is also increasing – the line on the graph slopes up from left to right.

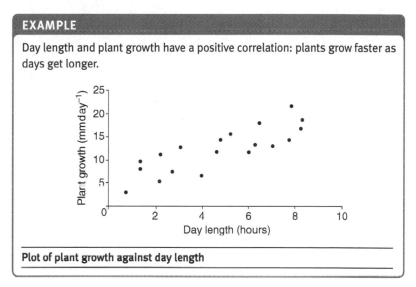
A **negative** correlation coefficient means that as the value of one variable goes up, the value for the other variable goes down – the graph slopes down from left to right.

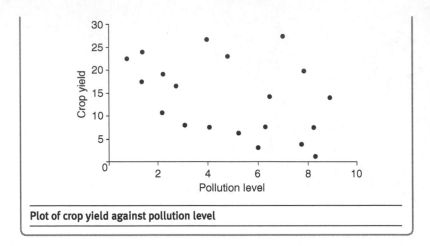

Plot of crop yield against pollution level

44.2 The correlation coefficient

The strength of a correlation is measured by the **correlation coefficient.**

The correlation coefficient is usually denoted by the Greek letter "ρ" (rho) for a population, for example $\rho = 0.8$, but by "r" for a sample, for example the **Pearson correlation coefficient.**

If there is a perfect relationship between the two variables then $\rho = 1$ (a positive correlation) or $\rho = -1$ (a negative correlation). If there is no correlation at all (the points on the graph are randomly scattered) then $\rho = 0$.

Interpreting the size of the correlation coefficient is subjective but the following may be a useful rule of thumb:

$\rho = 0$–0.2   very low and probably meaningless

$\rho = 0.2$–0.4   a low correlation that might warrant further investigation

$\rho = 0.4$–0.6   a reasonable correlation

$\rho = 0.6$–0.8   a high correlation

$\rho = 0.8$–1.0   a very high correlation. Possibly too high! Check for errors or other reasons for such a high correlation

This guide also applies to negative correlations.

> **EXAMPLE**
>
> With the day-length and plant-growth data above, $r = 0.8$, indicating a high correlation.
>
> However, the correlation between crop growth and pollution levels is lower: $r = -0.39$.

44.3 Calculating the Pearson correlation coefficient

The most often used measure of correlation is the **Pearson correlation coefficient**. To calculate this, first we need the means for both sets of data:

$$\bar{x} \text{ and } \bar{y}$$

For each value of x and y we then calculate:

$$x - \bar{x} \text{ and } y - \bar{y}$$

These are used for the correlation coefficient formula:

$$r = \frac{\Sigma (x - \bar{x})(y - \bar{y})}{\sqrt{\Sigma (x - \bar{x})^2 \; \Sigma (y - \bar{y})^2}}$$

EXAMPLE

A medical biochemist is interested in how closely blood glucose levels in humans relate to HbA1c (a measure of how much glucose is attached to haemoglobin molecules). The values, and their means, in a sample of eight people with diabetes mellitus are given in the table below.

Table of blood glucose and HbA1c in eight patients with diabetes mellitus		
Patient	Blood glucose $(mmol\,l^{-1})$	HbA1c (%)
A	5.1	5.8
B	4.6	6.9
C	6.3	8.3
D	8.3	6.1
E	9.7	7.8
F	12.0	8.4
G	12.7	10.8
H	14.1	9.1
Totals	72.8	63.2
Mean	9.1 $(\bar{y})$	7.9 $(\bar{x})$

Comparing the pairs of measurements gives the following graph:

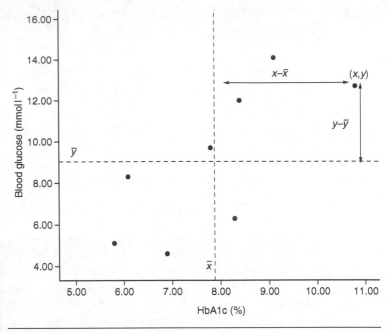

Plot of blood glucose against HbA1c

Also shown on the graph are dotted vertical and horizontal lines to indicate the mean values, $\bar{x}$ and $\bar{y}$, and the arrows show $x - \bar{x}$ and $y - \bar{y}$ for one of the patients.

Taking the values of x and y from the previous table, we can create a new data table to help solve the equation:

Table showing calculated values for $x - \bar{x}$, $y - \bar{y}$, $(x - \bar{x})^2$, and $(y - \bar{y})^2$

Patient	$x - \bar{x}$	$y - \bar{y}$	$(x - \bar{x})^2$	$(y - \bar{y})^2$
A	−2.1	−4.0	4.41	16.00
B	−1.0	−4.5	1.00	20.25
C	0.4	−2.8	0.16	7.84
D	−1.8	−0.8	3.24	0.64
E	−0.1	0.6	0.01	0.36
F	0.5	2.9	0.25	8.41
G	2.9	3.6	8.41	12.96
H	1.2	5.0	1.44	25.00

Substituting these values into the correlation coefficient formula gives the following result:

$$r = \frac{\Sigma\,(x - \bar{x})(y - \bar{y})}{\sqrt{\Sigma\,(x - \bar{x})^2\,\Sigma(y - \bar{y})^2}} = \frac{31.05}{\sqrt{91.46 \times 18.92}} = 0.7464$$

Thus $r = 0.75$, suggesting a high correlation between blood glucose levels and HbA1c.

44.4 Limitations of correlation

Correlation tells us about the strength of the association between the variables, but doesn't tell us about cause and effect in that relationship.

Take care when interpreting the significance of correlations. If a correlation is significant, we also need to consider the size of the correlation. If a study is sufficiently large, even a small correlation will have a high level of significance.

Also, bear in mind that a correlation only tells us about linear (straight-line) relationships between variables. Two variables may be strongly related but not have a straight line relationship, giving a low correlation coefficient.

Test yourself

The answers are given on page 221.

Question 44.1
The data comparing height and femur length for ten men are as follows:

Height (cm)	177	165.5	179.25	171.75	169	173	166.25	168	170.75	164
Femur length (cm)	106.25	100.5	111	107	100	118.25	108.5	100.25	105.25	90

Calculate the correlation coefficient between height and femur length to two significant figures.

Question 44.2
A researcher measures the mean germination
rates of delphinium seeds, *Delphinium
cardinale*, planted at different times after
harvesting.
Calculate the correlation coefficient, *r*.

	Table of germination rates of *D. cardinale* seeds at different times after harvesting	
	Germination rate (%)	Time after harvesting seeds (months)
	60	0
	53	4
	52	8
	38	12
	32	16
Mean	47	8

 # Regression

Regression analysis is used to quantify how one set of data relates to another. It is used when one of the variables is dependent on the other, independent, variable.

45.1 Linear regression

Linear regression is used where there is a linear (straight-line) relationship between the variables.

> **EXAMPLE**
>
> In humans, HbA1c is a measure of how much glucose is attached to haemoglobin molecules. It is dependent on the blood glucose level.
>
> The graph in Section 44.3 shows that the relationship is linear, so we can use linear regression to analyse it.

45.2 The line of best fit

In Section 23.6, we learnt how the **line of best fit** drawn on a scatter plot is the straight line that best shows the trend of the plotted points.

From that we can calculate the gradient. If we continue the line of best fit until it crosses the y-axis, we can also estimate the constant c for the formula of a straight-line graph, $y = mx + c$, explained in Chapter 9.

Regression analysis is the process of calculating this formula mathematically.

45.3 The regression line

A **regression line** is the line of best fit through the data points on a graph.

The **regression coefficient** gives the *gradient* of the graph, in that it gives the change in value of one outcome, per unit change in the other.

The **regression constant** gives the *position* of the line on the graph – it is the point where the line crosses the vertical axis.

So, the formula for the regression line is:

$$y = mx + c$$

where x is the independent variable, y is the dependent variable, m is the regression coefficient, and c is the regression constant.

45.4 Calculating the line of best fit

To fit a regression line to a scatter plot, we need the minimum vertical distance (deviation, d) from each point to the line.

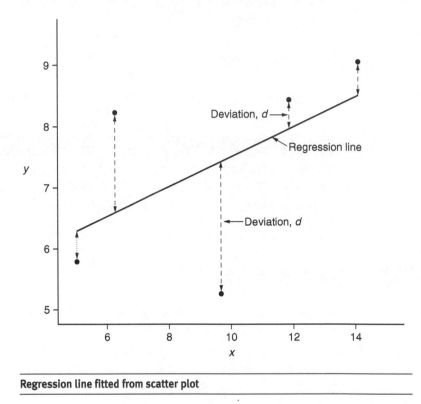

Regression line fitted from scatter plot

In this graph, the sum of the deviations above the line equals the sum of the deviations below the line.

For a straight-line graph, termed **linear regression**, this is usually calculated by the **method of least squares**, a calculation designed to give the smallest possible sum of the *squares* of the deviations.

Calculating the regression coefficient uses the formula:

$$m = \frac{\sum(x - \bar{x})(y - \bar{y})}{\sum(x - \bar{x})^2}$$

The regression line always goes through the mean values of x and y, $\bar{x}$ and $\bar{y}$. We can therefore substitute $\bar{x}$, $\bar{y}$ and the regression coefficient, m, into the regression line formula

$$\bar{y} = m\bar{x} + c$$

and calculate the regression constant, c.

Once we know the regression constant and coefficient, we can calculate the value of y for any given value of x.

EXAMPLE

We want to quantify how altitude affects the diversity of tree and shrub species.

We find eight equal sized areas of woodland, of similar ages, that have had similar forestry management. The numbers of tree and shrub species at different altitudes are given below.

Table of number of tree and shrub species at different altitudes		
Quadrat	Altitude (m)	Number of tree and shrub species
A	40	58
B	90	55
C	150	33
D	160	46
E	250	31
F	360	29
G	420	16
H	610	4
Mean	260	34

Substituting these values into the formula for the regression coefficient gives the following result:

$$m = \frac{\Sigma(x - \bar{x})(y - \bar{y})}{\Sigma(x - \bar{x})^2} = -\frac{23790}{257600} = -0.0924$$

Substituting into the regression line formula $\bar{y} = m\bar{x} + c$ gives:

$$34 = (-0.0924 \times 260) + c$$

so, $c = 58.0$

The graph for this regression is shown here:

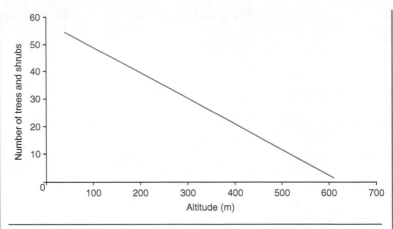

Regression graph of number of tree and shrub species at different altitudes

To predict the number of trees and shrubs for a given altitude we can now plug it into its regression formula $y = -0.0924x + 58$.

So, if we want to know the likely number of species at 300 m,

Number of species $= y = (-0.0924 \times 300) + 58 \approx 30$

45.5 Other values used with regression

We may wish to calculate the **standard error** of an estimate for a regression coefficient and constant. This indicates the accuracy that can be given to the calculations.

If there is considerable scatter, we may also wish to calculate the **significance level** of the regression – the probability that the calculated gradient is significantly different to zero.

Another useful value is the R^2 **value**. This shows how much a change in the dependent variable depends on a change in the independent variable.

> **EXAMPLE**
>
> In the example above:
> - the standard error of the estimate is 6.10;
> - it is significant to $P < 0.001$ and it is therefore highly likely that the gradient is different to zero;
> - the R^2 value is 0.91, so 91% of the variation in number of species is accounted for by variation in altitude.

45.6 Other types of regression

So far we have discussed linear regression, where the line that best fits the points is straight. Many biological relationships give curved graphs. These can often be "transformed" to a straight-line relationship, for instance by taking the logarithm of a set of data.

Other forms of regression include logistic regression and Poisson regression.

Logistic regression is used where each case in the sample can only belong to one of two groups (e.g. having disease or not) with the outcome as the probability that a case belongs to one group rather than the other.

Poisson regression is mainly used to study time between rare events.

45.7 Watch out for …

Regression should not be used to make predictions outside the range of the original data. In the example above, we can only make predictions from altitudes which are between 40 and 610 m.

45.8 Regression or correlation?

Regression and correlation are easily confused.

Correlation measures the *strength* of an association between variables.

Regression *quantifies* an association. It should only be used if one of the variables is thought to precede or cause the other.

Test yourself

The answers are given on pages 221–23.

Question 45.1
Pierce (1948) *mechanically* measured the frequency of chirps (or pulses of sound) made by a striped ground cricket at various ground temperatures.

Chirps/second	20	16	19.8	18.4	17.1	15.5	14.7	15.7
Temperature (°F)	88.6	71.6	93.3	84.3	80.6	75.2	69.7	71.6

Chirps/second	15.4	16.3	15	17.2	16	17	14.4
Temperature (°F)	69.4	83.3	79.6	82.6	80.6	83.5	76.3

a) Calculate the linear regression coefficient and constant for the number of chirps at a given temperature.

b) Produce a linear regression graph.
c) Use your results to estimate the number of chirps per second at a temperature of 90°F.

Question 45.2

Large warm-blooded animals have lower resting heart rates than small ones.
Turn the data into a linear regression by taking the log of each value, then calculate the regression coefficient and constant.

Use the results to calculate the expected resting heart rate of a warm-blooded animal of mass 15 kg.

Table comparing body mass and resting heart rate for different species		
	Mass (kg)	Resting heart rate (beats min^{-1})
Mouse	0.02	700
Rat	0.2	400
Cat	5	150
Dog	10	120
Man	70	70
Horse	450	40

46 Bayesian statistics

Bayesian analysis is a totally different statistical approach to the classical, "frequentist" statistics explained in this book.

It is being used increasingly often, for example in structural biology.

46.1 Prior and posterior distributions

In Bayesian statistics, rather than considering the sample of data on its own, a **prior distribution** is set up using information that is already available. For instance, a researcher may give a numerical value and weighting to previous opinion and experience as well as previous research findings.

One consideration is that different researchers may put different weighting on the same previous findings.

The new sample data are then used to adjust this prior information to form a **posterior distribution**. Thus these resulting figures have taken *both* the disparate old data *and* the new data into account.

Answers to "test yourself" questions

Answer 2.1
The factors of 18 are 1, 2, 3, 6, 9 and 18.
The factors of 21 are 1, 3, 7 and 21.
The factors of 24 are 1, 2, 3, 4, 6, 8, 12 and 24.
The highest common factor is 3.

Answer 2.2
$7(4 + 3)(5 - 2) = 7 \times 7 \times 3 = 147$

Answer 2.3
Calculating the sum in the brackets, we need to do the division before the addition, giving:
$16(4) - 10 \div 5$
Then the division and multiplication give:
$64 - 2$
So the sum works out to:
62

Answer 2.4
21 has 4 factors, 1, 3, 7 and 21, so it is *not* a prime number.
22 has 4 factors, 1, 2, 11 and 22, so it is *not* a prime number.
23 has only 2 factors, 1 and 23, so it *is* a prime number.

Answer 2.5
7 squared $= 7^2 = 7 \times 7 = 49 \, \text{m}^2$

Answer 2.6
$8 \times 8 = 64$, so the square root of $64 = \sqrt{64} = 8$
The sample is 8 by 8 mm wide.

Answer 2.7
$15 \div 0.6 = 25$, so the patch needs to be 25 cm^2
$\sqrt{25} = 5$, so the patch will be 5×5 cm.

Answer 2.8
40 by 40 by 40 mm cubed $= 40^3 = 40 \times 40 \times 40$
$= 64\,000 \, \text{mm}^3$

Answer 2.9
$3.4 \times 3.4 \times 3.4 = 39.304$, so the boxes are approximately 39 nm^3.

Answer 2.10
$4 \times 4 \times 4 = 64$, so the cube root of $64 = \sqrt[3]{64} = 4$
The sample is 4 by 4 by 4 mm.

Answer 3.1
$\frac{5}{6} \times 72 = 60$

Answer 3.2
20 and 24 have a common factor of 4, so
$\frac{20}{24} = \frac{20 \div 4}{20 \div 4} = \frac{5}{6}$

Answer 3.3
The reciprocal of $\frac{24}{28}$ is $\frac{28}{24}$, or $1\frac{4}{24}$. This can be simplified to $1\frac{1}{6}$.

Answer 3.4
$\frac{2}{5} \times \frac{9}{10} = \frac{2 \times 9}{5 \times 10} = \frac{18}{50}$
This can be simplified to $\frac{9}{25}$.

Answer 3.5
$\frac{2}{5} \div \frac{9}{10} = \frac{2}{5} \times \frac{10}{9} = \frac{2 \times 10}{5 \times 9} = \frac{20}{45} = \frac{4}{9}$

Answer 3.6
Here, the lowest common denominator is 14.
$\frac{6 \times 2}{7 \times 2} + \frac{9}{14} = \frac{12}{14} + \frac{9}{14} = \frac{21}{14} = 1\frac{7}{14} = 1\frac{1}{2}$

Answer 3.7
Convert $1\frac{3}{8}$ completely to a fraction: $\frac{11}{8}$
The lowest common denominator of 12 and 8 is 24.
$\frac{11 \times 3}{8 \times 3} - \frac{7 \times 2}{12 \times 2} = \frac{33}{24} - \frac{14}{24} = \frac{33 - 14}{24} = \frac{19}{24}$

Answer 3.8
$\frac{5}{8} = 0.625$, so $1\frac{5}{8} = 1.625$

Answer 3.9
$\frac{1}{5} \times \frac{1}{3} = \frac{1}{5 \times 3} = \frac{1}{15}$

Answer 3.10
a) $\frac{430}{4400}$ which can be simplified to $\frac{43}{440}$
b) $\frac{3960}{4400} = 0.9$

Answer 3.11
a) $\frac{7}{24}$
b) $\frac{2}{7}$
c) $\frac{7}{24} \times 80 = 23\frac{1}{3}$ years = 23 years and 4 months

Answer 4.1
$\frac{43.2}{72} \times 100 = 60\%$

Answer 4.2

40% of $375 = \left(\dfrac{40}{100}\right) \times 375 = 150$

So, 150 g of the original sample was water.

Answer 4.3

The peak flow rate has increased by 160 litres per minute.

$\dfrac{160}{400} \times 100 = 40\%$

So, it has increased *by* 40%. However, note that it has increased *to* 140% of its original (the original rate of 100%, plus the growth of 40%).

Answer 4.4

$\dfrac{129}{120} = 1.075$ – the proportional increase is therefore 0.075

$0.075 \times 100 = 7.5$ – so there has been a 7.5% increase in the patients' waist sizes.

Answer 4.5

A reduction of 18% is the same as 82% of the original (100% − 18% = 82%).

82% of 900 g is $0.82 \times 900 = 738$ g.

Answer 4.6

The cell concentration has increased by 888 million cells per ml, which is

$\dfrac{888 \text{ million}}{24 \text{ million}}$ of the original concentration.

Percentage increase $= \dfrac{888}{24} \times 100 = 3700\%$.

Answer 4.7

a) $\dfrac{75}{100} \times 1\,600\,000 = 1\,200\,000$ or 1.2 million

b) $1\,200\,000 \times \dfrac{1}{100} = 12\,000$

c) $\dfrac{12\,000}{80} \times 20 = 3000$

Answer 4.8

$\dfrac{0.25}{0.5} \times 100 = 50$ – so the patient needs 50 ml

Answer 4.9

Year 1: $150 \times \dfrac{(100 - 20)}{100} = 120$ kg

Year 2: $120 \times \dfrac{(100 + 20)}{100} = 144$ kg

Answer 5.1

Moving the decimal point four places to the right gives 10 to the power of −4, so the measurement is 4.5×10^{-4} m.

Answer 5.2

Moving the decimal point nine places to the right gives 3 000 000 000 base pairs.

Answer 5.3

Vitamin A: 800 µg $= 800 \times 10^{-6}$ g (standard form) $= 0.0008$ g (ordinary form).

Vitamin D: 5 µg $= 5 \times 10^{-6}$ g $= 0.000005$ g.

Vitamin E: 12 mg $= 1.2 \times 10^{-2}$ g $= 0.012$ g.

The highest dose is vitamin E and the lowest dose is vitamin D.

Answer 5.4

$150 = 1.5 \times 10^2$ people per km^2

40 by 40 km $= 1600$ km$^2 = 1.6 \times 10^3$ km^2

$(1.5 \times 10^2)(1.6 \times 10^3) = (1.5 \times 1.6)(10^{2+3}) = 2.4 \times 10^5$ people

So we would expect a population of 2.4×10^5 people in the 40 by 40 km area.

Answer 5.5

The phytoplankton are $0.015 \times 10^{-3} = (1.5 \times 10^{-2}) \times 10^{-3} = 1.5 \times 10^{-2-3} = 1.5 \times 10^{-5}$ m long.

The whales are 3×10 m.

$\dfrac{3 \times 10}{1.5 \times 10^{-5}} = 2 \times 10^{(+1+5)} = 2 \times 10^6$

So the whales are 2×10^6 or 2 000 000 (2 million) times longer than the phytoplankton.

Answer 6.1

a) A drop of antiseptic forms a circle of radius 1 mm. Area $= \pi r^2 = \pi \cdot 1^2 = 3.1416$ mm^2

3 drops give an area of $3\pi = 9.42$ mm^2, to three significant figures.

b) The volume (V) of the cylinder of skin treated $= A \cdot h$ where A is the area of skin covered and h is the depth of skin penetrated.

So, $V = A \cdot h = 9.42 \times 0.5 = 4.71$ mm^3.

Answer 6.2

The volume of a sphere is $= \dfrac{4}{3} \pi r^3$.

The yolk diameter is 24 mm, so its radius is 12 mm.

$V = \dfrac{4}{3} 3.1416(12^3) = 7238$ mm^3 to the nearest whole number.

Answer 7.1

1.050 m is stated to four significant figures.

Answer 7.2

$58.44 \div 0.137 = 426.569$ gm^{-3}.

The least precise value is the volume of water, which is given to three significant figures. So, the concentration may only be given to three significant figures:

427 gm^{-3}

Answer 7.3

The error is ± 0.5 g, so the mass could be anywhere between 55.5 and just below 56.5 g.

Answer 7.4

$14.9 + 1.3 = 16.2$ g dl^{-1}

$14.9 - 1.3 = 13.6$ g dl^{-1}

So we can be confident that the patient's haemoglobin level is between 13.6 and 16.2 g dl^{-1} and so within the normal range.

Answer 7.5

a) 2.956×10^{-7} m

b) 0.0000002956 m

Answer 8.1

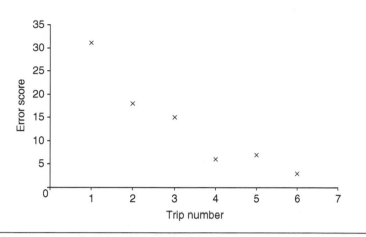

Plot of number of errors made by a rat in a maze

Answer 8.2

$R = K(N_A)$

$K = \dfrac{R}{N_A} = \dfrac{8}{6} = 1.3\dot{3}$

Answer 8.3

The increase in risk can be stated by the equation:

$$\dfrac{\text{Radiation}}{\text{dose}} \approx \dfrac{1 \text{ additional cancer case}}{1000 \text{ patients}}$$

A thousandth of that radiation dose gives this equation:

$$\dfrac{\text{Radiation dose}}{1000} \approx \dfrac{1 \text{ additional cancer case}}{1000 \times 1000 \text{ patients}}$$

So, the lower dose can be predicted to produce one extra case in every million people exposed.

Answer 9.1

You could have used any portion of the graph. For example, when the x value changes from 10 to 50, the y value changes from 5000 to 25 000.

$\text{Gradient} = \dfrac{y_2 - y_1}{x_2 - x_1} = \dfrac{25\,000 - 5000}{50 - 10} = 500\,\text{m}\,\text{min}^{-1}$

$500\,\text{m}\,\text{min}^{-1} = 500 \times 60\,\text{m}\,\text{hour}^{-1} = 30\,000\,\text{m}\,\text{hour}^{-1} = 30\,\text{km}\,\text{hour}^{-1}$.

Answer 9.2

The formula for a straight line graph is:

$y = mx + c$

If we substitute in:

for y — L = length of baby in millimetres
for m — rate of growth = 10 mm per week
for x — A = age in weeks
for c — length at birth = 500 mm

Then the equation for the girl's growth is:

$L = 10A + 500$

Note how in the graph we concertina the y-axis between 0 and 490. This allows us to focus on the part of the graph that we are interested in.

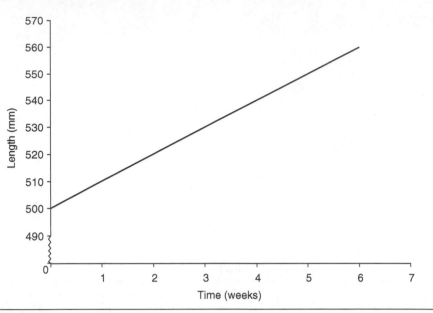

Graph of growth of a baby girl

Answer 9.3
First calculate the gradient. From the end of the 1st to the end of the 3rd year it has grown from 200 to 400 mm, so
Gradient $= \frac{y_2 - y_1}{x_2 - x_1} = \frac{400 - 200}{3 - 1} = 100$
Thus the equation for the line is $y = 100x + c$.
Substituting in the height at year 1, we get
$200 = (100 \times 1) + c$
So, $c = 200 - (100 \times 1) = 100$
The sapling was 100 mm high when planted.

Answer 9.4
$y = mx + c$, so
% increase of tissue nitrogen =
$\frac{\text{change in algal uptake of nitrogen}}{\text{change in time}} \times$
time exposed + constant $= \frac{0.9 - 0.2}{6} \times$ time
exposed $+ 0.2 = 0.1166\, t + 0.2$

Answer 9.5
a) $y = mx + c$, so
Height $= \frac{\text{change in height}}{\text{change in nitrogen rate}} \times$ nitrogen rate
+ constant $= \frac{50 - 30}{250} \times$ nitrogen rate + 30 =
0.08 (nitrogen rate) + 30
b) Height $= (0.08 \times 100) + 30 = 38$ cm

Answer 10.1
$27a^5 + 2a^3 + 5a^2 + 7a$
This can be further simplified to
$a(27a^4 + 2a^2 + 5a + 7)$

Answer 10.2
$\frac{a^2 b}{a^3} \times \frac{a^4 b^2}{b^3} = \frac{a^{2+4} b^{1+2}}{a^3 b^3} = \frac{a^6 b^3}{a^3 b^3} = a^{6-3} b^{3-3} =$
$a^3 b^0 = a^3$

Answer 10.3
Both top and bottom of the fraction can be divided by (i.e. have a common factor of) $a^2 b^3$ $(c + 2d)$
$\frac{a^3 b^3 (c + 2d)^4}{a^2 b^4 (c + 2d)} = \frac{a(c + 2d)^3}{b}$

Answer 10.4
1) No
2) Yes
$\frac{c^4 d^2 + b^2 c^2 d^2}{c^2 + b^2} = \frac{c^2 d^2 (c^2 + b^2)}{c^2 + b^2} = c^2 d^2$
3) No

Answer 10.5
$5a(2a - b^2) = (5a \times 2a) - (5a \times b^2) = 10a^2 - 5ab^2$

Answer 10.6
The highest common factor is $3a^2 b^2$.
Pull this out and the remainder is $2a + 3b^2$.
The answer is therefore $3a^2 b^2 (2a + 3b^2)$.

Answer 10.7
$a^2 - 4b^2 = (a + 2b)(a - 2b)$

Answer 10.8
a) $V = \pi r^2 L$
b) Measure the length of the bone and measure the diameter; halve the diameter to find the radius.

c) Long bone density $= \dfrac{m}{\pi r^2 l}$

d) g cm^{-3}

Answer 10.9

Squaring the first equation gives $v^2 = (u + at) \cdot (u + at) - u^2 + 2uat + (at)^2 = u^2 + 2uat + a^2 t^2$

So $u^2 + 2as = u^2 + 2uat + a^2 t^2$

Subtracting u^2 from each side of the equation gives

$2as = 2uat + a^2 t^2$

Dividing each side by $2a$ gives

$s = ut + \dfrac{1}{2} at^2$

Answer 11.1

The highest power is 5, so it is a 5th degree polynomial.

Answer 11.2

$$2x^5 + 7x^4 + 5x^3 \qquad\qquad +4$$
$$\underline{- (6x^4) - (9x^3) - (-x^2) - (5)}$$
$$= 2x^5 + \quad x^4 - 4x^3 \quad +x^2 \;\; -1$$

Answer 11.3

$$+\,8x^9 \qquad\quad +12x^6 \qquad\qquad +24x^4$$
$$\quad -2x^7 \qquad\qquad\qquad -3x^4 - 6x^2$$
$$\qquad\qquad\qquad +10x^5 \qquad\quad +15x^2 +30$$
$$\overline{8x^9 - 2x^7 + 12x^6 + 10x^5 + 21x^4 + 9x^2 + 30}$$

Answer 11.4

$x^2 - 6x + 9$ can be factorised to $x^2 - 2(3x) + 3^2$.

When $a = 3$, this is directly equivalent to the 2nd polynomial in the table in Section 11.5:

$x^2 - 2xa + a^2$

which the table shows us factorises to

$(x - a)^2$

Thus,

$x^2 - 6x + 9$ factorises to $(x - 3)^2$.

You can check this by multiplying out

$(x - 3)(x - 3)$.

Answer 12.1

$x^2 = \dfrac{12}{3} = 4$

$x = \sqrt{4} = 2$

Answer 12.2

Subtracting 1 from each side gives:

$4y^3 = x - 1$

Divide both sides by 4:

$y^3 = \dfrac{x - 1}{4}$

Taking the cube root of each side gives the solution:

$y = \sqrt[3]{\dfrac{x - 1}{4}}$

Answer 12.3

a) $a = \dfrac{v - u}{t}$

b) $a = \dfrac{40 - 0}{0.5} = 80$ m s^{-2}

c) $0.3 = (0 \times t) + \dfrac{1}{2} \times 80 \times t^2$

$0.3 = 40 \times t^2$

$t^2 = 7.5 \times 10^{-3}$

$t = 0.087$ seconds.

Answer 12.4

a) $V_E = V_T \times F$

b) $V_E = V_A + V_D$

c) $V_{CO_2} = V_A \times F_A CO_2$

d) $V_{CO_2} = V_A \times k \times P_a CO_2$

e) $V_A = 0.863\, V_{CO_2} \times P_a CO_2{}^{-1}$

Answer 13.1

When factorised, this becomes

$(x + 4)(x + 2) = 0$

To set the first factor to zero, x must be -4.

To set the second factor to zero, x must be -2.

Answer 13.2

$$x = \dfrac{-b \pm \sqrt{b^2 - 4ac}}{2a} = \dfrac{-6 \pm \sqrt{6^2 - (4 \times 1 \times 8)}}{2 \times 1}$$

$$= \dfrac{-6 \pm \sqrt{36 - (32)}}{2} = \dfrac{-6 \pm \sqrt{4}}{2} = \dfrac{-6 \pm 2}{2}$$

So the two solutions are:

$\dfrac{-6 + 2}{2} = -2$ and $\dfrac{-6 - 2}{2} = -4$

Answer 13.3

Putting the x^2 and x terms on one side and the constant on the other we get:

$x^2 + 6x = -8$

The coefficient (multiplier) of x^2 is 1, and dividing both sides of an equation by 1 leaves it exactly the same.

We now take half the coefficient of x, i.e. 3 (half of 6), square it (giving 9), and add it to both sides.

$x^2 + 6x + 9 = -8 + 9 = 1$

Factorising the left side:

$(x + 3)^2 = 1$

Taking the square root of both sides gives:

$\sqrt{(x + 3)^2} = \sqrt{1} = \pm 1$

Thus $x + 3 = \pm 1$

This means that:

$x = +1 - 3 = -2$ and $x = -1 - 3 = -4$

Answer 13.4

$10 = 5t + \dfrac{1}{2}(10)\, t^2$

$5t^2 + 5t - 10 = 0$

So $t^2 + t - 2 = 0$

$(t + 2)(t - 1) = 0$

So $t = 1$ or -2

It is impossible for the float to take -2 seconds to fall, so it must take 1 second to fall the 10 m height of the waterfall.

Answer 14.1

If $2x + y = 8$, then

$y = 8 - 2x$

Substituting this value of y into $3x + 2y = 14$ gives:

$3x + 2(8 - 2x) = 3x - 4x + 16 = -x + 16 = 14$

So $x = 2$

Putting this value of x into $2x + y = 8$ gives:

$4 + y = 8$

Thus $y = 4$
Therefore, $x = 2$, $y = 4$.

Answer 14.2

$2x + y = 8$　　(1)
and
$3x + 2y = 14$　　(2)
We need to multiply both sides of equation (1)
by 2:
$2(2x + y) = 2 \times 8$
Multiplying this out gives:
$4x + 2y = 16$
Both equations can now be manipulated to
equal $2y$.
Equation (1) becomes:
$2y = 16 - 4x$
Equation (2) becomes:
$2y = 14 - 3x$

Now that both equations are equal,
$16 - 4x = 2y = 14 - 3x$
so, $16 - 4x = 14 - 3x$
This simplifies to: $x = 2$
Substituting into either of the original equations
gives: $y = 4$
Therefore, $x = 2$, $y = 4$.

Answer 14.3

The equation that describes the population of
the first village is:
$y = 50x + 1000$
where y = population size, and x = time in years.
The population of the second village can be
described by:
$y = -50x + 1600$
The graph of these equations is shown below:

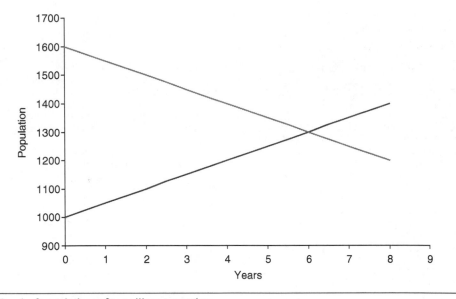

Graph of populations of two villages over time

The intersection is at 6 years, when the
population will be 1300.
To solve $y = 50x + 1000$ and $y = -50x + 1600$ by
elimination, as both equations equal y, we can
write:
$50x + 1000 = -50x + 1600$
This can be manipulated to:
$50x + 50x = 1600 - 1000$
So,
$100x = 600$
$x = 6$
So, the population size will be the same in 6
years.
Substituting this value of x into $y = 50x + 1000$
gives:
$y = (50 \times 6) + 1000 = 1300$

So in 6 years both villages will have 1300
inhabitants.

Answer 14.4

a) Between 12 and 25 m depth
b) No. $y = -\frac{1}{4}(75) + 25 = 6.25$ m, so 75% of
surface light intensity is only available above
6.25 m; however, the nutrients are only present
below 12 m.
c) $12 = -\frac{1}{4}x + 25$
$\frac{1}{4}x = 13$
Minimum percentage of surface light intensity =
52%.

Answer 15.1
In the equation $a_n = a + (n-1)d$,
$a = 12$ as the first count is 12; $d = 4$ as the common difference is 4; $n = 10$.
Thus we get
$12 + (10-1)4 = 12 + 36 = 48$
In the 10th year, 48 robins would be expected.

Answer 15.2
Again, $a = 12$, $d = 4$ and $n = 10$.
Putting these into the formula
$$S_n = \frac{n}{2}[2a + (n-1)d]$$
gives:
$$S_{10} = \frac{10}{2}[(2 \times 12) + (10-1)4] = 5 \times (24 + 36) = 300$$

Answer 15.3
a) Arithmetic.
b) $a_n = 1986 + (n-1)76$.
c) $a_2 = 1986 + (2-1)76 = 1986 + 76 = 2062$
$a_3 = 1986 + (3-1)76 = 1986 + 152 = 2138$

Answer 15.4
a) Number of cells (n) in each ring $= 6n$
b) Arithmetic
c) $S_n = \frac{n}{2}[2 \times 6a + (n-1)6] + 1 = \frac{n}{2}[6 + 6n] + 1$
$= 3n^2 + 3n + 1$, where n is the number of rings.
Note that the term +1 has been added so that the single first cell is included in the equation.
d) $S_{14} = 3 \times 14^2 + 3 \times 14 + 1 = 631$. So, the sum of the rings up to and including the 14th ring is 631.

Answer 15.5
The first term is 250, so $a = 250$. The common ratio is 3, so $r = 3$, and $n = 7$.
Substituting these into the formula $a\,r^{n-1}$ gives
$250 \times 3^{7-1} = 250 \times 3^6 = 182\,250$

Answer 15.6
Once again, $a = 250$, $r = 3$, and $n = 7$.
Putting these into the formula
$$S_n = \frac{a(r^n - 1)}{r - 1} \text{ gives:}$$
$$S_n = \frac{250(3^7 - 1)}{3 - 1} = \frac{250 \times 2186}{2} = 273\,250 \text{ plants}$$

Answer 15.7
a) Geometric
b) The nth term is given by ar^{n-1}
$a = 1.241$, $r = 1.014$ so the nth term $= 1.241 \times 1.014^{(n-1)}$
c) 2013 population $= 1.241 \times 1.014^2 = 1.276$ billion

Answer 16.1
$\frac{(a+2)^7}{(a+2)^5} = (a+2)^{7-5} = (a+2)^2 = a^2 + 4a + 4$

Answer 16.2
$\sqrt[3]{(2a-1)^6} = (2a-1)^{6/3} = (2a-1)^2 = 4a^2 - 4a + 1$

Answer 16.3
$38 = \frac{124}{h^2}$

Answer 16.3 (continued)
$h^2 = \frac{124}{38} = 3.263$
$h = \sqrt{3.263} = 1.81$ m to three significant figures.

Answer 16.4
$Q \propto r^4$
So doubling the radius gives 16 (because $2^4 = 16$) times the flow rate.

Answer 16.5
The light travels 3×10^4 m in 0.0001 second,
$\frac{3 \times 10^4 \text{ m}}{3.4 \times 10^2 \text{ ms}^{-1}} = 0.88 \times 10^2 \text{s} = 88 \text{ seconds.}$

Answer 16.6
$\frac{64 \times 2^{30} \text{ bytes}}{1.44 \times 2^{20} \text{ bytes}} = 46\,000$ to two significant figures.

Answer 17.1
1) $\log(1.2 \times 10^{-5}) = \log 0.000012 = -4.92$
$pH = -\log[H^+] = -(-4.92) = 4.92$
So the pH is 4.9 to two significant figures.
2) $pH = -\log[H^+] = 6.3$
$\log[H^+] = -6.3$
$[H^+] = 5.0 \times 10^{-7}$

Answer 17.2
$\ln e^4 = 4$

Answer 17.3
$10^7 = 10^4 \times 2^n$
So, $2^n = \frac{10^7}{10^4} = 10^3 = 1000$
$n \approx 9.97$
$G = \frac{3}{9.97} \approx 0.301$ hours, which is 18 minutes to two significant figures.

Answer 18.1
Initial number of patients $= N_0 = 5$; 3 weeks later, i.e. at $t = 3$ weeks, there are 25 patients, so $N_3 = 25$.
Plugging these into the growth–decay formula, $N = N_0 e^{kt}$, we get:
$25 = 5e^{3k}$
This manipulates to:
$\frac{25}{5} = 5 = e^{3k}$
$\ln 5 = 3k$
$\frac{\ln 5}{3} = k$
$k = 0.54$ per week

Answer 18.2
Using our calculated infectivity growth constant of $k = 0.54$ per week, we can calculate how many patients would be affected in another 4 weeks, i.e. 7 weeks after initial diagnosis ($t = 7$ weeks). $N_0 = 5$ as before.
$N_7 = N_0 e^{kt} = 5e^{0.54 \times 7} = 5e^{3.78} = 219$ patients

Answer 18.3
$N = 2N_0$ and $t = 20$ minutes
Growth–decay formula $N = N_0 e^{kt}$
$2N_0 = N_0 e^{kt}$

$\dfrac{2N_0}{N_0} = e^{kt}$

$2 = e^{kt}$

$\ln 2 = kt \ln e$, but $\ln e = 1$, so $\ln 2 = kt$

$t = 20$ minutes

So $k = \dfrac{\ln 2}{20} \approx 0.0347 \approx 3.47 \times 10^{-2}$

Answer 18.4

$k = 7.7 \times 10^{-4}$ per minute $= 0.0462$ per hour

$N_{72} = 1 \times e^{0.0462 \times 72} \approx e^{3.3264} \approx 28$

So, 28 mycobacteria would be present.

Answer 18.5

We can use any initial value for N_0, say $N_0 = 1$.
After 8 days ($t = 8$ days), we have half the initial value, $N_8 = 0.5$.
Substituting these values into $N = N_0 \, e^{kt}$, we get

$0.5 = 1e^{8k}$

Taking the natural logarithm of both sides gives:

$\ln 0.5 = 8k$

$\dfrac{\ln 0.5}{8} = k$

$k = -0.087$ per day

Answer 18.6

$N = N_0 \, e^{kt}$

$N = \dfrac{N_0}{2}$

$\dfrac{1}{2} = 1e^{(-9.5 \times 10^{-3} \times t)}$

$\ln \dfrac{1}{2} = (-9.5 \times 10^{-3}) \, t \ln e$

But $\ln e = 1$

So $t = \dfrac{\ln \frac{1}{2}}{-9.5 \times 10^{-3}} \approx 73$ hours.

Answer 19.1

The rate of increase in heat loss at that point will be:

$\dfrac{dy}{dx} = 2(50x) = 2 \times 50 \times 1.2 = 120 \, \mathrm{W\,m^{-1}}$

So, at 1.2 m in length, the heat loss is increasing at the rate of 120 W m^{-1} growth.

Answer 19.2

When $x = 30$, the rate of increase, or gradient, is

$\dfrac{dy}{dx} = \dfrac{3x^2}{60} = \dfrac{x^2}{20} = \dfrac{30^2}{20} = 45$

Thus, when the radius of the colony reaches 30 mm, the rate of increase is 45 wasps per mm increase in radius.

Answer 19.3

The differential of $y = 3x^{20} - 8$ is

$\dfrac{dy}{dx} = 20(3x^{20-1}) - 0 = 60x^{19}$

Answer 19.4

$y = \dfrac{3}{x^5}$ is the same as $y = 3x^{-5}$.

So,

$\dfrac{dy}{dx} = 3(-5x^{-5-1}) = -15x^{-6}$.

Answer 20.1

a) Let $t = 3x^2 + 2x + 5$

So $y = t^2$, $\dfrac{dy}{dt} = 2t$ and $\dfrac{dt}{dx} = 6x + 2$

So using the chain rule:

$\dfrac{dy}{dx} = \dfrac{dy}{dt} \cdot \dfrac{dt}{dx} = 2(3x^2 + 2x + 5) \cdot (6x + 2)$

$= (6x^2 + 4x + 10) \cdot (6x + 2) = 36x^3 + 24x^2 + 60x + 12x^2 + 8x + 20 = 36x^3 + 36x^2 + 68x + 20 = 4(9x^3 + 9x^2 + 17x + 5)$

b) In this equation y is the product of two functions, so we need to use the product rule.

Let $u = x^3$, then $\dfrac{du}{dx} = 3x^2$

Let $t = x - 2$, then $\dfrac{dt}{dx} = 1$

Let $v = (x - 2)^2$, then $v = t^2$

So $\dfrac{dv}{dt} = 2t = 2(x - 2) = 2x - 4$

Using the chain rule:

$\dfrac{dv}{dx} = \dfrac{dv}{dt} \cdot \dfrac{dt}{dx} = (2x - 4) \cdot 1 = 2x - 4$

Plugging these into the product rule gives

$\dfrac{dy}{dx} = u \cdot \dfrac{dv}{dx} + v \cdot \dfrac{du}{dx} = x^3 \cdot (2x - 4) + (x - 2)^2 \cdot 3x^2$

$= 2x^4 - 4x^3 + 3x^2(x^2 - 4x + 4) = 2x^4 - 4x^3 + 3x^4 - 12x^3 + 12x^2 = 5x^4 - 16x^3 + 12x^2$

c) Here again, y is the product of two functions so we need to use the product rule.

Let $u = x^3$, then $\dfrac{du}{dx} = 3x^2$

Let $v = e^x$, then $\dfrac{dv}{dx} = e^x$

So using the product rule:

$\dfrac{dy}{dx} = u \cdot \dfrac{dv}{dx} + v \cdot \dfrac{du}{dx} = x^3 \cdot e^x + e^x \cdot 3x^2 = e^x(x^3 + 3x^2)$

d) Let $u = e^x$, then $\dfrac{du}{dx} = e^x$

Let $v = \ln x$, then $\dfrac{dv}{dx} = \dfrac{1}{x}$

So using the product rule:

$\dfrac{dy}{dx} = u \cdot \dfrac{dv}{dx} + v \cdot \dfrac{du}{dx} = e^x \cdot \dfrac{1}{x} + \ln x \cdot e^x = e^x \left(\dfrac{1}{x} + \ln x \right)$

e) Here, y is a function divided by another function so we need to use the quotient rule.

Let $u = x^2 - 1$, then $\dfrac{du}{dx} = 2x$

Let $v = x^2 + 1$, then $\dfrac{dv}{dx} = 2x$ and $v^2 = (x^2 + 1)^2$

So using the quotient rule:

$\dfrac{dy}{dx} = \dfrac{v \dfrac{du}{dx} - u \dfrac{dv}{dx}}{v^2} = \dfrac{(x^2 + 1) \cdot 2x - (x^2 - 1) \cdot 2x}{(x^2 + 1)^2} =$

$\dfrac{2x(x^2 + 1 - x^2 + 1)}{(x^2 + 1)^2} = \dfrac{4x}{(x^2 + 1)^2}$

Answer 20.2

a) Again, y is a function divided by another function so we need to use the quotient rule.

Let $u = x$, then $\frac{du}{dx} = 1$

Let $v = x^2 + 1$, then $v^2 = (x^2 + 1)^2$, and $\frac{dv}{dx} = 2x$

So by the quotient rule:

$$\frac{dy}{dx} = \frac{v \frac{du}{dx} - u \frac{dv}{dx}}{v^2} = \frac{(x^2 + 1)\cdot 1 - x\cdot 2x}{(x^2 + 1)^2} =$$

$$\frac{x^2 + 1 - 2x^2}{(x^2 + 1)^2} = \frac{1 - x^2}{(x^2 + 1)^2}$$

b) $\frac{dy}{dx} = \frac{1 - x^2}{(x^2 + 1)^2} = 0$

This can be written as $(1 - x^2)(x^2 + 1)^{-2} = 0$

So $1 - x^2 = 0$ and $x^2 = 1$, therefore $x = \pm 1$

Answer 20.3

a) If height is in metres, then BSA $= \left(\frac{100h \times w}{3600}\right)^{\frac{1}{2}}$

BMI $= \frac{w}{h^2}$ so $w =$ BMI $\cdot h^2$

Therefore BSA $= \left(\frac{100h \times \text{BMI} \cdot h^2}{3600}\right)^{\frac{1}{2}} = \left(\frac{\text{BMI} \cdot h^3}{36}\right)^{\frac{1}{2}}$

b) To find the change of body surface area with respect to height, we need to differentiate using the chain rule.

Let $t = \frac{ch^3}{36}$, then $\frac{dt}{dh} = \frac{3ch^2}{36} = \frac{ch^2}{12}$

BSA $= t^{\frac{1}{2}}$ so $\frac{d(\text{BSA})}{dt} = \frac{1}{2}t^{-\frac{1}{2}} = \frac{1}{2} \cdot \frac{1}{t^{\frac{1}{2}}} =$

$\frac{1}{2}\sqrt{\left(\frac{36}{ch^3}\right)} = \frac{1}{2} \cdot \frac{6}{c^{\frac{1}{2}} \cdot h^{\frac{3}{2}}} = 3c^{-\frac{1}{2}} h^{-\frac{3}{2}}$

$\frac{d(\text{BSA})}{dh} = \frac{dt}{dh} \cdot \frac{d(\text{BSA})}{dt} = \frac{ch^2}{12} \cdot 3c^{-\frac{1}{2}} h^{-\frac{3}{2}} =$

$\frac{c^{\frac{1}{2}} h^{\frac{1}{2}}}{4} = \sqrt{\left(\frac{ch}{16}\right)}$

For our young man, we plug in $h = 1.55$ m (his height halfway through the year) and $c = 24$ (his BMI, which is $\frac{54}{1.5^2} = 24$ kg m^{-2}) into this equation, giving:

$\frac{d(\text{BSA})}{dh} = \sqrt{\left(\frac{ch}{16}\right)} = \sqrt{\left(\frac{24 \times 1.55}{16}\right)} = 1.524 \, \frac{\text{m}^2}{\text{m}}$

to four significant figures.

So halfway through the year, for every 1 cm increase in his height, his body surface area was increasing by 0.015 m^2.

Answer 21.1

$\int 7x^4 \, dx = 7 \frac{x^{4+1}}{4+1} + C = \frac{7x^5}{5} + C$

The integral is thus $\frac{7x^5}{5} + C$

Answer 21.2

Subsituting $n = 3$ into the formula

$\int x^n \, dx = \frac{x^{n+1}}{n+1} + C$

we get the definite integral

$A = \int_3^5 2x^3 \, dx = \left[\frac{2x^4}{4} + C\right]_3^5$

$= \left(\frac{2 \times 5^4}{4} + C\right) - \left(\frac{2 \times 3^4}{4} + C\right)$

$= (312.5 + C) - (40.5 + C) = 272.$

Answer 21.3

$\int_0^4 f(t) \, dt = -\frac{1}{4} \cdot \frac{1}{4}t^4 + \frac{3}{2} \cdot \frac{1}{3}t^3 + t + c =$

$\left[-\frac{1}{16}t^4 + \frac{1}{2}t^3 + t + c\right]_0^4 = (-16 + 32 + 4 + c) -$

$(+ c) = 20$

Therefore the person metabolises 20 calories in 4 minutes.

Answer 22.1

a) $u = x^2 + 1$, so $\frac{du}{dx} = 2x$, therefore $x = \frac{1}{2} \cdot \frac{du}{dx}$

Hence $\int x(x^2 + 1)^3 \, dx = \int \frac{1}{2} \cdot \frac{du}{dx} \cdot u^3 dx =$

$\int \frac{1}{2} \cdot u^3 \cdot \frac{dx}{du} \cdot \frac{du}{dx} \, du = \int \frac{1}{2} \cdot u^3 \cdot 1 \, du = \frac{1}{8}u^4 + c$

Substituting back in for u gives $\frac{1}{8}(x^2 + 1)^4 + c$

b) $u = x^2$ so $\frac{du}{dx} = 2x$, therefore $x = \frac{1}{2} \cdot \frac{du}{dx}$

Hence $\int xe^{x^2} \, dx = \int \frac{1}{2} \cdot \frac{du}{dx} \cdot e^u \, dx = \int \frac{1}{2} e^u du =$

$\frac{1}{2}e^u + c$

Substitute back in for u:

$\frac{1}{2}e^u + c = \frac{1}{2}e^{x^2} + c$

c) $u = x^4$

So $\frac{du}{dx} = 4x^3$, and $x^3 = \frac{1}{4} \cdot \frac{du}{dx}$

$\int x^3 \sqrt{(x^4 - 1)} \, dx = \int \frac{1}{4} \cdot \frac{du}{dx} \cdot \sqrt{(u - 1)} \, dx =$

$\int \frac{1}{4}(u - 1)^{\frac{1}{2}} \, du$

$\int \frac{1}{4}(u - 1)^{\frac{1}{2}} \, du = \frac{3}{2} \cdot \frac{1}{4}(u - 1)^{\frac{3}{2}} + c =$

$\frac{3}{8}(u - 1)^{\frac{3}{2}} + c$

Substitute back in for u:

$\frac{3}{8}(u - 1)^{\frac{3}{2}} + c = \frac{3}{8}(x^4 - 1)^{\frac{3}{2}} + c$

d) $u = 1 + 2x^2$

So $\frac{du}{dx} = 4x$, and $x = \frac{1}{4} \cdot \frac{du}{dx}$

When $x = 1$, $u = 1 + 2(1)^2 = 3$

When $x = 2$, $u = 1 + 2(2)^2 = 9$

So $\int_1^2 x(1 + 2x^2) \, dx = \int_1^2 \frac{1}{4} \cdot \frac{du}{dx} \cdot u \, dx =$

$\int_3^9 \frac{1}{4} \cdot u \, du = \left[\frac{1}{8}u^2 + c\right]_3^9$

$= \left(\frac{1}{8} \cdot 9^2 + c\right) - \left(\frac{1}{8} \cdot 3^2 + c\right) =$

$10.125 + c - 1.125 - c = 9$

Answer 22.2

Let $v = x^2$, and $\frac{dv}{dx} = 2x$

$\frac{du}{dx} = e^x$, and $u = e^x$

$\int v \frac{du}{dx} dx = uv - \int u \frac{dv}{dx} dx$

$\int x^2 e^x dx = e^x \cdot x^2 - \int e^x \cdot 2x \, dx$

To find

$\int 2x \, e^x \, dx$

Let $v = 2x$, and $\frac{dv}{dx} = 2$

$\frac{du}{dx} = e^x$, and $u = e^x$

$\int v \frac{du}{dx} dx = uv - \int u \frac{dv}{dx} dx$

$\int 2x \, e^x \, dx = 2x \, e^x - \int 2e^x \, dx = 2x \, e^x - 2e^x$

Substitute this back into the first equation

$\int x^2 e^x dx = e^x \cdot x^2 - \int e^x \cdot 2x \, dx = x^2 e^x - 2xe^x - 2e^x$
$= e^x(x^2 - 2x - 2) + c$

Answer 22.3

It is easier to differentiate $\ln x$ and therefore define v as $\ln x$.

Let $v = \ln x$, and $\frac{dv}{dx} = \frac{1}{x}$

$\frac{du}{dx} = x^5$, and $u = \frac{1}{6} x^6$

$\int v \frac{du}{dx} dx = uv - \int u \frac{dv}{dx} dx$

$\int_1^2 x^5 \ln x \, dx = \frac{1}{6} x^6 \cdot \ln x - \int_1^2 \frac{1}{6} x^6 \cdot \frac{1}{x} dx =$

$\frac{1}{6} x^6 \cdot \ln x - \int_1^2 \frac{1}{6} x^5 dx$

$\left[\frac{1}{6} x^6 \cdot \ln x - \frac{1}{36} x^6 + c \right]_1^2 = \left(\frac{32}{3} \ln 2 - \frac{16}{9} + c \right) - \left(0 - \frac{1}{36} + c \right)$

$\frac{32}{3} \ln 2 - \frac{16}{9} + \frac{1}{36} = \frac{32}{3} \ln 2 - \frac{7}{4} = 5.644$ to four significant figures.

Answer 22.4

For 0–2 hours: $\int_0^2 4x \, dx = \left[2x^2 + c \right]_0^2 = 8 - 0 = 8$

For 2–12 hours: $\int_2^{12} 8e^{\left(\frac{2}{5} - \frac{x}{5}\right)} dx$

Let $u = \frac{2}{5} - \frac{x}{5}$ then $x = 2 - 5u$, so $\frac{dx}{du} = -5$

$\int_2^{12} 8e^{\left(\frac{2}{5} - \frac{x}{5}\right)} dx = \int_2^{12} 8e^u \cdot \frac{dx}{du} dx$

When $x = 2$, $u = 0$.
When $x = 12$, $u = -2$

$\int_0^{-2} 8e^u \cdot -5 \, du = -40 \int_0^{-2} e^u \, du = -40 \left[e^u + c \right]_0^{-2} =$

$-40(e^{-2} - e^0) = -40(0.135 - 1) = 34.6$
So the total area under the curve is $34.6 + 8 = 42.6$ mg h l^{-1}.

Answer 25.1
$5.5 \, pg = 5.5 \times 10^{-15} \, kg$

Answer 25.2
$(5 \times 10^3)(6 \times 10^3) = 30 \times 10^6 = 3 \times 10^7 \, m^2$.

Answer 25.3
$0°C = 273.15$ K so
$-40.5°C = -40.5 + 273.15 = 232.65$ K
The temperature in °C was given to one decimal place, so the temperature in K also needs to be given to one decimal place:
$-40.5°C = 232.7$ K.

Answer 25.4
a) An angstrom is 10^{-10} m, so 4000 Å = 4000×10^{-10} m = 4×10^{-7} m in standard form.
b) 4000×10^{-10} m = 400×10^{-9} m = 400 nm using a metric prefix.

Answer 26.1

C	$1 \times 12.01 =$	12.01 Da
O_2	$2 \times 16.00 =$	32.00 Da

The molecular mass of $CO_2 = 44.01$ Da

Answer 26.2
Number of moles $= \dfrac{\text{mass}}{\text{relative molecular mass}} =$
$\dfrac{528.36}{176.12} = 3 \, mol$

Answer 26.3

C_5	$5 \times 12.01 =$	60.05 Da
H_4	$4 \times 1.01 =$	4.04 Da
N_4	$4 \times 14.01 =$	56.04 Da
O_3	$3 \times 16.00 =$	48.00 Da

Molecular mass of uric acid = 168.13 Da.

Answer 26.4
$M_r = 300$, so 1.5 kg contains $\dfrac{1.5 \times 10^3 (g)}{300} = 5$ moles.

A mole of any substance contains 6.022×10^{23} molecules, so 5 moles contains $5 \times 6.022 \times 10^{23}$
$= 30.11 \times 10^{23} = 3.033 \times 10^{24}$ molecules.

Answer 26.5
$\dfrac{\text{mass in human body}}{\text{molecular mass}} = \dfrac{250}{507.18} = 0.49$ mol to two significant figures.

Answer 26.6
Mass of glucose = $180.18 \times 0.5 \, M \times 0.2 \, l = 18.018$ g

Answer 26.7
Mass of urea = $60.06 \times 8 \, M \times 0.1 \, l = 48.048$ g

Answer 26.8
$\dfrac{0.25 \, M}{1 \, M} = \dfrac{500 \, ml}{x \, ml}$
$x = \dfrac{1 \times 500}{0.25} = 2000 \, ml$
So, 2 l of 0.25 M solution can be prepared.

Answer 26.9

$$\frac{\text{required molarity of new solution}}{\text{molarity of stock solution}} =$$
$$\frac{\text{required volume of stock solution}}{\text{required total volume of new solution}}$$

Plugging in the values gives $\frac{5\,M}{8\,M} =$

$$\frac{100ml}{\text{required total volume of new solution (ml)}}$$

Rearranging the equation, required total volume

of new solution $= \frac{8 \times 100}{5} = 160$ ml

So, the total final volume needs to be 160 ml. To achieve that, we need to add 60 ml of water to the 100 ml of stock 8 M urea solution.

Answer 26.10

A 5% w/v solution of glucose contains 5 g per 100 ml, which is 50 g l^{-1}.

50 g $l^{-1} = \frac{50}{180.18}$ M $= 0.28$ M to two significant figures.

Answer 26.11

A 40% w/v solution of urea contains 40 g per 100 ml, which is 400 g l^{-1}.

400 g $l^{-1} = \frac{400}{60.06}$ M $= 6.66$ M to three significant figures.

Answer 26.12

The molarity of the solution $= \frac{\text{no. of moles}}{\text{volume (l)}} =$

$$\frac{\text{mass(g)}}{\text{relative molecular mass} \times \text{solution volume (l)}} =$$

$$\frac{20}{40 \times 1} = 0.5 \text{ M}.$$

As NaOH has a single replaceable OH$^-$ ion per molecule its normality is the same as its molarity, in this case 0.5 N.

Answer 27.1

In 0.1 M HCl, as there is almost complete dissociation, the concentration of H$^+$ is effectively also 0.1 M, so [H$^+$] = 0.1
pH $= -\log[H^+] = -\log 0.1 = 1$
The pH of 0.1 M HCl is 1.

Answer 27.2

pH $= -\log[H^+] = 2$
So, $\log[H^+] = -2$
[H$^+$] = 0.01 M

Answer 27.3

[H$^+$][OH$^-$] = 10^{-14}
So [H$^+$][OH$^-$] = [H$^+$] × 0.1 = 10^{-14}
Thus [H$^+$] = $\frac{10^{-14}}{0.1}$ = 10^{-13}
pH $= -\log[H^+] = -\log 10^{-13} = 13$
So the pH of 0.1 M NaOH is 13.

Answer 27.4

pH $= -\log[H^+] = 8.5$
So, $\log[H^+] = -8.5$
[H$^+$] = 3.16×10^{-9} M to three significant figures.

Putting this into the equation for the ion product of water:
[H$^+$][OH$^-$] = $(3.16 \times 10^{-9}) \times$ [OH$^-$] = 10^{-14}
[OH$^-$] = $\frac{10^{-14}}{3.6 \times 10^{-9}}$ = 2.78×10^{-6} to three significant figures.

Answer 27.5

$pK_a = -\log K_a = 9.25$
so
$\log K_a = -9.25$
$K_a = 5.62 \times 10^{-10}$.

Answer 27.6

Sodium hydroxide is almost completely ionised to OH$^-$ ions in water, so 1 ml of 1 M solution contributes 0.001 moles of OH$^-$ ions to a total volume of 1001 ml (1 litre plus the 1 ml NaOH that has been added).

[OH$^-$] = $\frac{0.001 \text{ moles}}{1.001 \text{ litre}}$ = 0.000999, which is 1.0×10^{-3} M to two significant figures.
Putting that into the equation for the ion product of water gives:
[H$^+$][OH$^-$] = 10^{-14} = [H$^+$] × 10^{-3}
[H$^+$] = $\frac{10^{-14}}{10^{-3}}$ = 10^{-14+3} = 10^{-11}
Converting that to pH:
pH $= -\log[H^+] = -\log 10^{-11} = 11.0$.

Answer 28.1

First calculate the resulting volume and then the molarities. The resulting volume would be 500 ml.
So, the resulting molarity of Tris–HCl would be:
$\frac{300}{500} = 0.6$ M.
The molarity of Tris base would be:
$\frac{200}{500} = 0.4$ M.

pH $= pK_a + \log \frac{[A^-]}{[HA]} = 8.3 + \log \frac{0.4}{0.6}$

$= 8.3 + (-0.1761) = 8.1239$

The pH is 8.1, to two significant figures.

Answer 28.2

pH $= pK_a + \log \frac{[\text{sodium acetate}]}{[\text{acetic acid}]} = 4.75 + \log$

$\frac{0.1}{0.2} = 4.75 + \log 0.5 = 4.75 - 0.301 = 4.45$ to three significant figures.

Answer 28.3

Desired pH = 7.4

$= 7.2 + \log \frac{[A^-]}{[HA]}$

So, $\log \frac{[A^-]}{[HA]} = 7.4 - 7.2 = 0.2$

$\frac{[A^-]}{[HA]} = 1.585$

We therefore need the ratio of 1 M conjugate base and diethylmalonic acid to be 1.585 to 1.

Amount of conjugate base needed $= \dfrac{1.585}{1+1.585} \approx$ 0.61 litres.

Amount of acid needed $= \dfrac{1}{1+1.585} \approx 0.39$ litres.

Answer 28.4
Converting the K_a to the pK_a:
$pK_a = -\log K_a = -\log (1.78 \times 10^{-5}) \approx 4.75$.
Plugging this into the Henderson–Hasselbalch
equation $pH = pK_a + \log \dfrac{[A^-]}{[HA]}$ gives:

$4.8 = 4.75 + \log \dfrac{[A^-]}{[HA]}$

$\log \dfrac{[A^-]}{[HA]} = 4.8 - 4.75 = 0.05$

$\dfrac{[A^-]}{[HA]} \approx 1.22$, so the ratio of conjugate base to
conjugate acid in the solution is 1.22 to three
significant figures.

Answer 29.1
The substrate concentration [S], of 1.5×10^{-5} M
is the same as the K_M for this enzyme and
substrate. We know that K_M is the substrate
concentration at which the reaction goes at half
V_{max} so:
$V = \dfrac{1}{2} \times 30 = 15$ µmol min^{-1}.

Answer 29.2
Putting the values into the Michaelis–Menten
equation gives:
$V = \dfrac{30 \times (1.5 \times 10^{-6})}{(1.5 \times 10^{-5}) + (1.5 \times 10^{-6})} = \dfrac{4.5 \times 10^{-5}}{1.65 \times 10^{-5}} =$
2.73 µmol min^{-1} to three significant figures.

Answer 29.3
The gradient of a Lineweaver–Burk plot is $\dfrac{K_M}{V_{max}}$

Plugging in the values gives $\dfrac{1}{V_{max}} = 0.5$

So, $V_{max} = 2$ µmol min^{-1}.

Answer 31.1
$\bar{x} = \dfrac{\Sigma x}{n} = \dfrac{11.7 + 11.9 + 12.2 + 12.7 + 13.0}{5} =$
$\dfrac{61.5}{5} = 12.3$ g dl^{-1}

Answer 31.2
The median is 12.5 g dl^{-1}, half-way between the
middle two values.

Answer 31.3
a) The sum of the observations is 1119. Dividing
by the 25 observations gives a mean of 44.76.
Ordering the observations from smallest to
largest, the middle observation (13th) gives the
median of 44.
The most commonly occurring observation gives
the mode of 32.
b) The mean is the most appropriate measure
because the data are evenly distributed around
the mean. The values for the mean and median
are very similar, as would be expected with
normally distributed data.

Answer 31.4
a) The data on seasons are best represented by
the mode because the data are categories but
the order is irrelevant to the analysis.
b) The data are continuous but not equally
distributed around a central value, so the
median value should be used.
c) Standardised tests should be equally
distributed around a central value and therefore
the mean is the most appropriate measure.

Answer 32.1
Answer 31.1 showed the calculation that gives
the sample mean, 12.3 g dl^{-1}.

Haemoglobin x	Sample mean $\bar{x}$	Deviation $x - \bar{x}$	Square of deviation $(x - \bar{x})^2$
11.7	12.3	−0.6	0.36
11.9	12.3	−0.4	0.16
12.2	12.3	−0.1	0.01
12.7	12.3	0.4	0.16
13.0	12.3	0.7	0.49
	Sum of deviations $\Sigma(x - \bar{x})^2$		1.18
	Dividing by $n-1$ gives the variance, V		$\dfrac{1.18}{(5-1)} = 0.295$
	The square root of the variance gives the SD		$\sqrt{0.295} = 0.543$

Thus SD $= 0.543$ g dl^{-1}.

Answer 32.2
This is a *sample* of lettuces, so we need to use
($n-1$) as the denominator.
The standard deviation is the square root of the
variance.

Variance $= \dfrac{\Sigma (x - \bar{x})^2}{n-1} =$

$\dfrac{(44-44.76)^2 + (45-44.76)^2 + (37-44.76)^2 + \ldots + (42-44.76)^2}{25-1}$

$= 53.85\dot{6}$

SD $= \sqrt{53.85\dot{6}} = 7.34$ to three significant
figures.

Answer 33.1
$z = \dfrac{(x - \mu)}{\sigma} = \dfrac{(15.5 - 12.5)}{1.2} = 2.5$

Answer 33.2
a) The z-score for the laboratory test is $\dfrac{(57-50)}{6}$
$= 1.16\dot{6}$

The z-score for the written test is $\dfrac{(64-50)}{14} = 1.0$

b) Since the z-score for the laboratory test is
larger than for the written test, the student did
better in the laboratory test than in the written
test compared to everyone else.

Answer 36.1
$SEM = \dfrac{SD}{\sqrt{n}} = \dfrac{0.4}{\sqrt{16}} = 0.1$

Answer 36.2

$$SEM = \frac{0.4}{\sqrt{36}} = 0.06\dot{6}$$

Answer 37.1

a) SEM $\times$ 1.96 = 0.8 $\times$ 1.96 = 1.57

$\bar{x}$ − 1.57 − 12.8 − 1.57 = 11.23

$\bar{x}$ + 1.57 = 12.8 + 1.57 = 14.37

so the 95% CI is 11.23 to 14.37

SEM $\times$ 2.58 = 0.8 $\times$ 2.58 = 2.06

$\bar{x}$ − 2.06 = 12.8 − 2.06 = 10.74

$\bar{x}$ + 2.06 = 12.8 + 2.06 = 14.86

so the 99% CI is 10.74 to 14.86

SEM $\times$ 3.29 = 0.8 $\times$ 3.29 = 2.63

$\bar{x}$ − 2.63 = 12.8 − 2.63 = 10.17

$\bar{x}$ + 2.63 = 12.8 + 2.63 = 15.43

so the 99.9% CI is 10.17 to 15.43

b) SD = SEM $\times$ $\sqrt{n}$ = 0.8 $\times$ $\sqrt{20}$ = 3.58 to three significant figures.

c) SEM = $\frac{SD}{\sqrt{n}}$ = $\frac{3.578}{\sqrt{200}}$ = 0.253 to three significant figures. $\bar{x}$ = 12.8.

95% CI = $\bar{x}$ ± (SEM $\times$ 1.96), so the 95% CI is 12.30–13.30 g dl^{-1}

99% CI = $\bar{x}$ ± (SEM $\times$ 2.58), so the 99% CI is 12.15–13.45 g dl^{-1}

99.9% CI = $\bar{x}$ ± (SEM $\times$ 3.29), so the 99.9% CI is 11.97–13.63 g dl^{-1}

Answer 38.1

The chance of throwing two sixes is:

- a 1 in 36 chance
- a 1/36 chance
- a 0.028 probability
- P = 0.028
- a 2.8% probability.

Answer 38.2

$\frac{1}{6} \times \frac{1}{6} \times \frac{1}{6} \times \frac{1}{6} \times \frac{1}{6} \times \frac{1}{6}$ = 1 in 46 656

Answer 39.1

The null hypothesis is that there is no difference between the effects of the two temperatures on the germination rate of the wheat seeds.

Answer 39.2

P = 0.25 means that the probability of the difference having happened by chance is 0.25 in 1, i.e. 1 in 4.

It is not statistically significant.

Answer 40.1

a) If the ages are normally distributed, use a parametric test.

b) Use a non-parametric test.

c) Because the data are categorical, a non-parametric test should be used. However, because there are many possible categories, some statisticians will be happy to use a parametric test.

Answer 41.1

$t = \frac{\bar{x} - E}{SEM} = \frac{62 - 70}{8} = -1$

The *t* distribution is symmetric, but usually only tabulated for positive values, so for $t = -1$, look up $t = +1$.

So, $t = 1$, df 11

The table in Appendix 2 shows that, for df 11, the critical value for 5% is 2.20. As $t < 2.20$, the result is not significant.

Answer 41.2

$t = \frac{d}{SE_d} = \frac{(36.9 - 36.8)}{0.04} = 2.5$

So, $t = 2.5$, df 19

Using the table of *t* values in Appendix 2, for 19 df the critical value of *t* for 5% is 2.09 and the critical value of *t* for 1% is 2.86. Our calculated value of 2.5 is therefore greater than needed for the 5% significance level but less than needed for the 1% level. Therefore the P value is less than 0.5 but greater than 0.01.

Answer 41.3

$t = \frac{\bar{x}_a - \bar{x}_b}{SE_d} = \frac{1.5 - 1.2}{0.1} = 3$

So, $t = 3$, df 58 (as an approximation use df = 60 from table)

Using the table of *t* values in Appendix 2, for 58 df the critical value of t for 1% is 2.66 and the critical value of *t* for 0.1% is 3.46. Our calculated value of 3.0 is therefore greater than needed for the 1% significance level but less than needed for the 0.1% level. Therefore the P value is less than 0.01 but greater than 0.001.

Answer 42.1

a) Gender is a significant variable as it has a P value of 0.039, i.e. <0.05.

Smoking status is not a significant variable as its P value is 0.397, i.e. >0.05.

b) The R^2 of 0.183 means that only 18.3% of the variation in FEV$_1$ is accounted for by gender and smoking.

Answer 43.1

The expected frequency, E, of cultures with colonies for the standard medium is given by:

E = total number of cultures with bacterial colony present $\times$ total number of standard culture Petri dishes/Total number of Petri dishes

So,

$E = \frac{304 \times 240}{480}$ = 152 for standard medium, colony present;

$E = \frac{176 \times 240}{480}$ = 88 for standard medium, no colony present;

$E = \frac{304 \times 240}{480}$ = 152 for new medium, colony present;

$E = \frac{176 \times 240}{480}$ = 88 for new medium, no colony present.

The following table shows the calculation of $\sum \frac{(O - E)^2}{E}$

Table calculating χ^2 for different culture media					
	Observed frequency, O	Expected frequency, E	$O - E$	$(O - E)^2$	$\dfrac{(O - E)^2}{E}$
Standard medium, colony	144	152	−8	64	0.4211
Standard medium, no colony	96	88	8	64	0.7273
New medium, colony	160	152	8	64	0.4211
New medium, no colony	80	88	−8	64	0.7273
				$\sum \dfrac{(O - E)^2}{E} = 2.297$	

So, $\chi^2 = 2.297$

Answer 43.2
df = (2 − 1)(2 − 1) = 1
The critical value for the 5% significance level is 3.84 for 1 df.
But in our example, χ^2 is only 2.297, so the result is not significant and the null hypothesis stands.

Answer 43.3

	Observed frequency, O	Expected frequency, E	$O - E$	$(O - E)^2$	$\dfrac{(O - E)^2}{E}$
Large spores, multiple outgrowth	80	65.3̇	14.6̇	215.1̇	3.293
Large spores, single outgrowth	40	54.6̇	−14.6̇	215.1̇	3.935
Small spores, multiple outgrowth	18	32.6̇	−14.6̇	215.1̇	6.585
Small spores, single outgrowth	42	27.3̇	14.6̇	215.1̇	7.870
				$\sum \dfrac{(O - E)^2}{E} = 21.683$	

So the chi-squared value is 21.7 to three significant figures.

df = (rows − 1) × (columns − 1) = 1 × 1 = 1

From Appendix 3, the critical chi-squared for 1 degree of freedom at the 0.1% level is 10.83. As our result of 21.7 is above this critical value, the result is highly significant with a $P<0.001$. So there is strong evidence of a relationship between the size of spores and whether the outgrowth is single or multiple.

Answer 44.1

x	y	$(x-\bar{x})$	$(y-\bar{y})$	$(x-\bar{x})^2$	$(y-\bar{y})^2$	$(x-\bar{x})\times$ $(y-\bar{y})$
Height	Femur length					
177	106.25	6.55	1.55	42.9025	2.4025	10.1525
165.5	100.5	−4.95	−4.2	24.5025	17.64	20.79
179.25	111	8.8	6.3	77.44	39.69	55.44
171.75	107	1.3	2.3	1.69	5.29	2.99
169	100	−1.45	−4.7	2.1025	22.09	6.815
173	118.25	2.55	13.55	6.5025	183.6025	34.5525
166.25	108.5	−4.2	3.8	17.64	14.44	−15.96
168	100.25	−2.45	−4.45	6.0025	19.8025	10.9025
170.75	105.25	0.3	0.55	0.09	0.3025	0.165
164	90	−6.45	−14.7	41.6025	216.09	94.815
Sum				220.475	521.35	220.6625

The coefficient of correlation for the variables is thus:

$$r = \frac{\Sigma(x-\bar{x})(y-\bar{y})}{\sqrt{\Sigma(x-\bar{x})^2\Sigma(y-\bar{y})^2}}$$

$$= \frac{220.6625}{\sqrt{220.475\times521.35}} = 0.65 \text{ to two}$$

significant figures.

Answer 44.2

$$r = \frac{\Sigma(x-\bar{x})(y-\bar{y})}{\sqrt{\Sigma(x-\bar{x})^2\Sigma(y-\bar{y})^2}}$$

$$= \frac{-284}{\sqrt{536\times160}} = -0.97$$

Answer 45.1

a)

	y	x	$(y-\bar{y})$	$(x-\bar{x})$	$(x-\bar{x})^2$	$(x-\bar{x})\times(y-\bar{y})$
	20	88.6	3.433	9.253	85.618	31.770
	16	71.6	−0.567	−7.747	60.016	4.390
	19.8	93.3	3.233	13.953	194.687	45.116
	18.4	84.3	1.833	4.953	24.532	9.081
	17.1	80.6	0.533	1.253	1.570	0.668
	15.5	75.2	−1.067	−4.147	17.198	4.423
	14.7	69.7	−1.867	−9.647	93.065	18.007
	15.7	71.6	−0.867	−7.747	60.016	6.714
	15.4	69.4	−1.167	−9.947	98.943	11.604
	16.3	83.3	−0.267	3.953	15.626	−1.054
	15	79.6	−1.567	0.253	0.064	−0.397
	17.2	82.6	0.633	3.253	10.582	2.060
	16	80.6	−0.567	1.253	1.570	−0.710
	17	83.5	0.433	4.153	17.247	1.800
	14.4	76.3	−2.167	−3.047	9.284	6.601
Sum	248.5	1190.2			690.017	140.073

$$m = \frac{\Sigma(x - \bar{x})(y - \bar{y})}{\Sigma(x - \bar{x})^2} = \frac{140.073}{690.017} = 0.203 \text{ to three}$$

significant figures.

$$\bar{y} = \frac{248.5}{15} = 16.5\dot{6}, \ \bar{x} = \frac{1190.2}{15} = 79.34\dot{6}$$

Substituting into the regression line formula

$\bar{y} = m\bar{x} + c$

$16.5\dot{6} = (0.203 \times 79.34\dot{6}) + c$

So, $c = 0.459$ to three significant figures.

Therefore the regression equation is:

$y = 0.203 \, x + 0.459$

b)

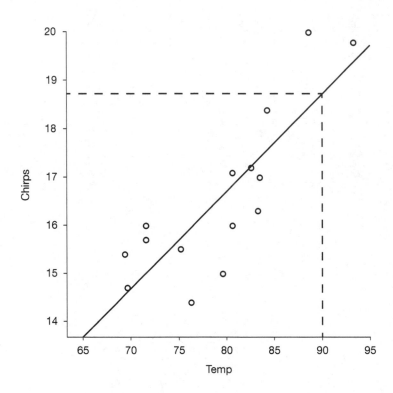

c) To predict the number of chirps for a given temperature we can put it into the regression formula: $y = 0.203 \, x + 0.459 = 0.203 \times 90 + 0.459 = 18.7$ chirps per second to three significant figures.

We can also estimate the value from the graph, as shown above.

Answer 45.2

	Mass (kg)	log of (mass)	Resting heart rate (beats min^{-1})	log of heart rate
Mouse	0.02	−1.70	700	2.85
Rat	0.2	−0.70	400	2.60
Cat	5	0.70	150	2.18
Dog	10	1.00	120	2.08
Man	70	1.85	70	1.85
Horse	450	2.65	40	1.60
Mean		0.63		2.19

After calculating the various values of $(x - \bar{x})$ and $(y - \bar{y})$ we can calculate:

$$m = \frac{\Sigma (x - \bar{x})(y - \bar{y})}{\Sigma (x - \bar{x})^2} = \frac{-3.72}{12.90} = -0.288$$

Substituting into the regression line formula $\bar{y} = m\bar{x} + c$ gives:

$2.19 = (-0.288 \times 0.63) + c$

Thus, $c = 2.37$

So, $y = -0.288x + 2.37$

For a 15-kg animal, $\log(15) = 1.176$

$y = (-0.288 \times 1.176) + 2.37 = 2.03$

This is a logarithm of the expected resting heart rate, so

Expected resting heart rate = $10^{2.03}$ = 107 beats min^{-1}

Appendix 1: Flow chart for choosing statistical tests

The key tests for each category are given in the lozenges.

The tests in *italics* are the non-parametric tests for each category.

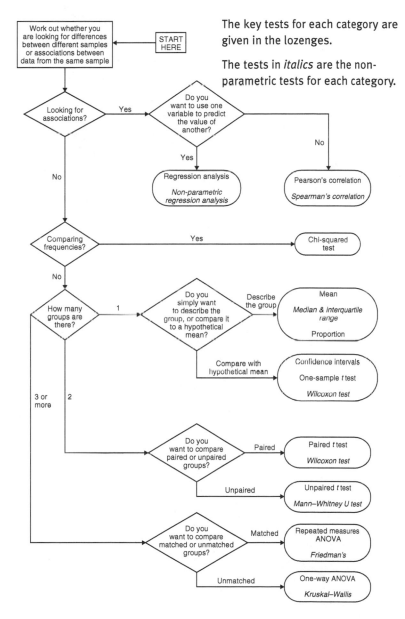

Work out whether you are looking for differences between different samples or associations between data from the same sample

START HERE

Looking for associations? — Yes → Do you want to use one variable to predict the value of another?

No — Looking for associations

Yes → Regression analysis / *Non-parametric regression analysis*

No → Pearson's correlation / *Spearman's correlation*

Comparing frequencies? — Yes → Chi-squared test

No

How many groups are there?

1 → Do you simply want to describe the group, or compare it to a hypothetical mean?

Describe the group → Mean / *Median & interquartile range* / Proportion

Compare with hypothetical mean → Confidence intervals / One-sample *t* test / *Wilcoxon test*

2 → Do you want to compare paired or unpaired groups?

Paired → Paired *t* test / *Wilcoxon test*

Unpaired → Unpaired *t* test / *Mann–Whitney U test*

3 or more → Do you want to compare matched or unmatched groups?

Matched → Repeated measures ANOVA / *Friedman's*

Unmatched → One-way ANOVA / *Kruskal–Wallis*

Appendix 2: Critical values for the *t* distribution

This table gives us the critical *t* values for different degrees of freedom and commonly used significance levels.

We normally need to use the two-tailed significance levels. See Section 41.2 for an explanation of one- and two-tailed testing.

We reject the null hypothesis if the calculated value of *t* is larger than the value on the table for a chosen significance level.

Degrees of freedom (df)	Significance level for two-tailed test		
	5%	1%	0.1%
	Significance level for one-tailed test		
	2.5%	0.5%	0.05%
1	12.71	63.66	636.58
2	4.30	9.92	31.60
3	3.18	5.84	12.92
4	2.78	4.60	8.61
5	2.57	4.03	6.87
6	2.45	3.71	5.96
7	2.36	3.50	5.41
8	2.31	3.36	5.04
9	2.26	3.25	4.78
10	2.23	3.17	4.59
11	2.20	3.11	4.44
12	2.18	3.05	4.32
13	2.16	3.01	4.22
14	2.14	2.98	4.14
15	2.13	2.95	4.07
16	2.12	2.92	4.01
17	2.11	2.90	3.97
18	2.10	2.88	3.92
19	2.09	2.86	3.88
20	2.09	2.85	3.85
25	2.06	2.79	3.73
30	2.04	2.75	3.65
40	2.02	2.70	3.55
50	2.01	2.68	3.50
60	2.00	2.66	3.46
70	1.99	2.65	3.43
80	1.99	2.64	3.42
90	1.99	2.63	3.40
100	1.98	2.63	3.39
Infinity	1.96	2.58	3.29

Note: A *t* distribution with infinite degrees of freedom is equivalent to the values of the normal distribution.

Appendix 3: Critical values for the chi-squared distribution

This table gives us the critical chi-squared values for different degrees of freedom and commonly used significance levels.

We reject the null hypothesis if our calculated value of chi-squared is larger than the value on the table for a chosen significance level.

Degrees of freedom (df)	Significance level		
	5%	1%	0.1%
1	3.84	6.63	10.83
2	5.99	9.21	13.82
3	7.81	11.34	16.27
4	9.49	13.28	18.47
5	11.07	15.09	20.51
6	12.59	16.81	22.46
7	14.07	18.48	24.32
8	15.51	20.09	26.12
9	16.92	21.67	27.88
10	18.31	23.21	29.59
11	19.68	24.73	31.26
12	21.03	26.22	32.91
13	22.36	27.69	34.53
14	23.68	29.14	36.12
15	25.00	30.58	37.70
16	26.30	32.00	39.25
17	27.59	33.41	40.79
18	28.87	34.81	42.31
19	30.14	36.19	43.82
20	31.41	37.57	45.31
25	37.65	44.31	52.62
30	43.77	50.89	59.70
40	55.76	63.69	73.40
50	67.50	76.15	86.66
60	79.08	88.38	99.61
70	90.53	100.43	112.32
80	101.88	112.33	124.84
90	113.15	124.12	137.21
100	124.34	135.81	149.45

Index

Where multiple page numbers are given, the bold page numbers indicate the key pages.